全国高等职业教育规划教材

网络设备配置与管理实训教程

主　编　危光辉　李　腾
副主编　罗小辉　郎登何
参　编　李贺华

机械工业出版社

本书适用于网络设备配置管理的实训课程，内容包括搭建实验环境，网络设备的基本配置，PC 与 DynamipsGUI 桥接，端口镜像与协议分析，静态路由，浮动静态路由，默认路由，RIPv1/v2、EIGRP、OSPF 的基本配置，向 RIP、EIGRP、OSPF 注入默认路由，RIPv2、EIGRP、OSPF 的验证，OSPF 的广播多路访问、重发布和路由归纳，多区域 OSPF、VLAN、STP、VTP 的配置，交换机的二层和三层链路聚合，基于时间和反射 ACL，路由器 DHCP 的配置，IPv6 的静态/动态路由和隧道配置，以及帧中继的多个配置实验等，既有基础性的实验，也有较深入实用的实验，其中部分实验在市场上的实训教材中不常见，但又非常实用，像端口镜像与协议分析、PC 桥接模拟软件等。本书编排结构上充分考虑教学以及自学的特点，每一个实验采用完整配置，结合及时注解以及归纳总结，既方便学习，又能拓展提高。

本书既讲解了基于真实硬件设备环境下的实验方法，又讲解了 Cisco Packet Tracer 和 DynamipsGUI 两种模拟软件的使用方法，从而使读者在完成本教程实验时，在没有硬件设备的条件下，只需一台计算机即可完成本书中的所有实验。

本书既可作为高职高专院校计算机及相关专业的实训教材，可配套《网络设备配置与管理》教材一起使用，也可单独作为有一定基础的企业网络技术人员的参考用书和职业培训教材，同时，还可作为 CCNA 考试的实验参考书。

图书在版编目（CIP）数据

网络设备配置与管理实训教程 / 危光辉，李腾主编. —北京：机械工业出版社，2016.2

全国高等职业教育规划教材

ISBN 978-7-111-52769-5

Ⅰ. ①网… Ⅱ. ①危… ②李… Ⅲ. ①网络设备—配置—高等职业教育—教材 Ⅳ. ①TP393

中国版本图书馆 CIP 数据核字（2016）第 017782 号

机械工业出版社（北京市百万庄大街22号 邮政编码100037）
策划编辑：鹿　征
责任编辑：张　帆
责任校对：张艳霞
责任印制：乔　宇
唐山丰电印务有限公司印刷
2016 年 2 月第 1 版·第 1 次印刷
184mm×260mm・13 印张・321 千字
0001—3000 册
标准书号：ISBN 978-7-111-52769-5
定价：32.00 元

凡购本书，如有缺页、倒页、脱页，由本社发行部调换

电话服务　　　　　　　　　网络服务
服务咨询热线：(010) 88379833　　机工官网：www.cmpbook.com
读者购书热线：(010) 88379649　　机工官博：weibo.com/cmp1952
　　　　　　　　　　　　　　　　教育服务网：www.cmpedu.com
封面无防伪标均为盗版　　　　　金　书　网：www.golden-book.com

出 版 说 明

《国务院关于加快发展现代职业教育的决定》指出：到 2020 年，形成适应发展需求、产教深度融合、中职高职衔接、职业教育与普通教育相互沟通，体现终身教育理念，具有中国特色、世界水平的现代职业教育体系，推进人才培养模式创新，坚持校企合作、工学结合，强化教学、学习、实训相融合的教育教学活动，推行项目教学、案例教学、工作过程导向教学等教学模式，引导社会力量参与教学过程，共同开发课程和教材等教育资源。机械工业出版社组织全国 60 余所职业院校（其中大部分是示范性院校和骨干院校）的骨干教师共同策划、编写并出版的"全国高等职业教育规划教材"系列丛书，已历经十余年的积淀和发展，今后将更加紧密结合国家职业教育文件精神，致力于建设符合现代职业教育教学需求的教材体系，打造充分适应现代职业教育教学模式的、体现工学结合特点的新型精品化教材。

"全国高等职业教育规划教材"涵盖计算机、电子和机电三个专业，目前在销教材 300 余种，其中"十五""十一五""十二五"累计获奖教材 60 余种，更有 4 种获得国家级精品教材。该系列教材依托于高职高专计算机、电子、机电三个专业编委会，充分体现职业院校教学改革和课程改革的需要，其内容和质量颇受授课教师的认可。

在系列教材策划和编写的过程中，主编院校通过编委会平台充分调研相关院校的专业课程体系，认真讨论课程教学大纲，积极听取相关专家意见，并融合教学中的实践经验，吸收职业教育改革成果，寻求企业合作，针对不同的课程性质采取差异化的编写策略。其中，核心基础课程的教材在保持扎实的理论基础的同时，增加实训和习题以及相关的多媒体配套资源；实践性较强的课程则强调理论与实训紧密结合，采用理实一体的编写模式；涉及实用技术的课程则在教材中引入了最新的知识、技术、工艺和方法，同时重视企业参与，吸纳来自企业的真实案例。此外，根据实际教学的需要对部分课程进行了整合和优化。

归纳起来，本系列教材具有以下特点：

1）围绕培养学生的职业技能这条主线来设计教材的结构、内容和形式。

2）合理安排基础知识和实践知识的比例。基础知识以"必需、够用"为度，强调专业技术应用能力的训练，适当增加实训环节。

3）符合高职学生的学习特点和认知规律。对基本理论和方法的论述容易理解、清晰简洁，多用图表来表达信息；增加相关技术在生产中的应用实例，引导学生主动学习。

4）教材内容紧随技术和经济的发展而更新，及时将新知识、新技术、新工艺和新案例等引入教材。同时注重吸收最新的教学理念，并积极支持新专业的教材建设。

5）注重立体化教材建设。通过主教材、电子教案、配套素材光盘、实训指导和习题及解答等教学资源的有机结合，提高教学服务水平，为高素质技能型人才的培养创造良好的条件。

由于我国高等职业教育改革和发展的速度很快，加之我们的水平和经验有限，因此在教材的编写和出版过程中难免出现问题和疏漏。我们恳请使用这套教材的师生及时向我们反馈质量信息，以利于我们今后不断提高教材的出版质量，为广大师生提供更多、更适用的教材。

机械工业出版社

全国高等职业教育规划教材计算机专业编委会成员名单

主　任　周智文

副主任　周岳山　林　东　王协瑞　张福强
　　　　　陶书中　眭碧霞　龚小勇　王　泰
　　　　　李宏达　赵佩华

委　员　（按姓氏笔画顺序）
　　　　　万　钢　万雅静　卫振林　马　伟
　　　　　马林艺　王兴宝　王德年　尹敬齐
　　　　　史宝会　宁　蒙　乔芃喆　刘本军
　　　　　刘剑昀　刘瑞新　刘新强　安　进
　　　　　李　强　杨　云　杨　莉　何万里
　　　　　余先锋　张洪斌　张瑞英　赵国玲
　　　　　赵海兰　赵增敏　胡国胜　钮文良
　　　　　贺　平　秦学礼　贾永江　顾正刚
　　　　　徐立新　唐乾林　陶　洪　黄能耿
　　　　　黄崇本　曹　毅　裴有柱

秘书长　胡毓坚

前　言

"实践出真知",只有通过实践,才能将理论知识转化为动手能力。在学习网络设备配置与管理的过程中,必须通过大量的实验,才能真正做到知识的融会贯通,进而培养和提升解决网络中遇到的实际问题的能力。本书作者结合了 11 年企业网络工程从业经验和 8 年多的"网络工程"课程教学经验,深切体会了企业对网络设备配置与管理的实际需求,并主持编写了这本实训教程。

本书既讲解了基于真实硬件设备环境下的实验方法,又讲解了 Cisco Packet Tracer 和 DynamipsGUI 两种模拟软件的使用方法,从而使读者在完成本教程实验时,在没有硬件设备的条件下,只需一台计算机即可完成本书中的所有实验。

一、本书特点

1. 内容安排

① 本书首先讲述从物理设备和模拟软件两个方面构建实训环境的方法,以确保后续实验能够顺利进行,实现学习的低成本和高效率。

② 实训内容从网络设备的配置基础开始,到最后实现与广域网的互联配置,按逻辑关系合理进行了组织。

③ 对实验过程中的关键配置及时注解,并对整个实验涉及的知识点进行归纳总结。

2. 编排形式

① 提出实验要求,即读者完成此实验后达到的实验目标。

② 编写实验说明,讲解本实验主要的内容、应用环境以及涉及的知识点。

③ 根据实验内容构建拓扑图。

④ 结合拓扑图给出完整配置,并及时对配置语句进行注解。

⑤ 最后是实验总结部分。本部分是实验调试的重要环节,也是整个实验的重点所在。

二、本书内容

本书适用于网络设备配置管理的实训课程,内容包括实验环境说明,网络设备的基本配置,PC 与 DynamipsGUI 桥接,端口镜像与协议分析,静态路由,浮动静态路由,默认路由,RIPv1/v2、EIGRP、OSPF 的基本配置,向 RIP、EIGRP、OSPF 注入默认路由,RIPv2、EIGRP、OSPF 的验证,OSPF 的广播多路访问、重发布和路由归纳,多区域 OSPF,VLAN、STP、VTP 的配置,交换机的二层和三层链路聚合、基于时间的 ACL 和反射 ACL,路由器 DHCP 的配置,IPv6 的静态/动态路由和隧道配置,以及帧中继的多个配置实验等,既有基础性的实验,也有较深入实用的实验,其中部分实验在市场上的实训教材中不常见，但又非常实用，如端口镜像与协议分析、PC 桥接模拟软件等。

三、致谢

本书由危光辉、李腾任主编，罗小辉、郎登何任副主编，李贺华参编。在本教程编写过程中，编者参考了国内外大量的计算机网络同类著作和文献，以及互联网上的相关资料，同时也得到了重庆电子工程职业学院计算机学院领导和同事们的大力支持，在此一并表示衷心的感谢！

由于作者水平有限，错漏之处在所难免，恳请广大读者批评指正。

编 者

目 录

出版说明
前言
实验环境说明 ··· 1
    一、用 console 口访问路由器 ··· 1
    二、通过 console 口连接路由器的操作步骤 ·· 2
    三、使用终端访问服务器的方式访问路由器 ·· 4
    四、Cisco Packet Tracer 模拟软件的使用方法 ·· 6
    五、DynamipsGUI 的使用方法 ·· 12
实验 1　IOS 基本命令 ·· 23
    一、实验要求 ·· 23
    二、实验说明 ·· 23
    三、实验拓扑 ·· 23
    四、实验过程 ·· 23
实验 2　路由器的基本操作 ·· 31
    一、实验要求 ·· 31
    二、实验说明 ·· 31
    三、实验拓扑 ·· 31
    四、实验过程 ·· 31
实验 3　将真实 PC 与 DynamipsGUI 桥接 ·· 34
    一、实验要求 ·· 34
    二、实验说明 ·· 34
    三、操作过程 ·· 34
实验 4　端口镜像与协议分析 ·· 42
    一、实验要求 ·· 42
    二、实验说明 ·· 42
    三、实验拓扑 ·· 42
    四、实验过程 ·· 42
实验 5　静态路由的配置 ··· 52
    一、实验要求 ·· 52
    二、实验说明 ·· 52
    三、实验拓扑 ·· 52
    四、配置过程 ·· 52
实验 6　浮动静态路由 ·· 57

一、实验要求	57
二、实验说明	57
三、实验拓扑	57
四、实验配置	57

实验 7　默认路由 60
　　一、实验要求 60
　　二、实验说明 60
　　三、实验拓扑 60
　　四、实验配置 60

实验 8　RIPv1 62
　　一、实验要求 62
　　二、实验说明 62
　　三、实验拓扑 62
　　四、实验配置 62

实验 9　RIPv2 66
　　一、实验要求 66
　　二、实验说明 66
　　三、实验拓扑 66
　　四、实验配置 67

实验 10　RIPv2 路由验证 71
　　一、实验要求 71
　　二、实验说明 71
　　三、实验拓扑 71
　　四、实验过程 71
　　五、实验总结 72

实验 11　向 RIP 注入默认路由 74
　　一、实验要求 74
　　二、实验说明 74
　　三、实验拓扑 74
　　四、实验过程 74
　　五、实验总结 76

实验 12　EIGRP 的配置 78
　　一、实验要求 78
　　二、实验说明 78
　　三、实验拓扑 78
　　四、实验配置 78
　　五、实验总结 79

实验 13　向 EIGRP 注入默认路由 85
　　一、实验要求 85

二、实验说明 ·· *85*
　　三、实验拓扑 ·· *85*
　　四、实验配置 ·· *85*
　　五、实验总结 ·· *87*
实验 14　EIGRP 验证 ·· *89*
　　一、实验要求 ·· *89*
　　二、实验说明 ·· *89*
　　三、实验拓扑 ·· *89*
　　四、实验配置 ·· *89*
　　五、实验总结 ·· *90*
实验 15　OSPF 的配置 ·· *91*
　　一、实验要求 ·· *91*
　　二、实验说明 ·· *91*
　　三、实验拓扑 ·· *91*
　　四、实验配置 ·· *91*
　　五、实验总结 ·· *92*
实验 16　OSPF 的广播多路访问 ·· *95*
　　一、实验要求 ·· *95*
　　二、实验说明 ·· *95*
　　三、实验拓扑 ·· *95*
　　四、配置过程 ·· *96*
　　五、实验总结 ·· *96*
实验 17　OSPF 默认路由 ·· *99*
　　一、实验要求 ·· *99*
　　二、实验说明 ·· *99*
　　三、实验拓扑 ·· *99*
　　四、实验过程 ·· *99*
　　五、实验总结 ·· *100*
实验 18　OSPF 验证 ·· *102*
　　一、实验要求 ·· *102*
　　二、实验说明 ·· *102*
　　三、实验拓扑 ·· *102*
　　四、实验过程及实验总结 ·· *102*
实验 19　多区域 OSPF 的配置 ·· *105*
　　一、实验要求 ·· *105*
　　二、实验说明 ·· *105*
　　三、实验拓扑 ·· *105*
　　四、实验过程及实验总结 ·· *106*
实验 20　VLAN 基本配置 ·· *112*

一、实验要求 ·· 112
　　二、实验说明 ·· 112
　　三、实验拓扑 ·· 112
　　四、配置过程 ·· 112
　　五、实验总结 ·· 115
实验 21　VLAN 间路由 ·· 117
　　一、实验要求 ·· 117
　　二、实验说明 ·· 117
　　三、基于三层交换机的 VLAN 间路由 ·· 117
　　四、基于路由器物理接口的 VLAN 间路由 ·· 119
　　五、单臂路由 ·· 120
实验 22　VTP 的配置 ·· 122
　　一、实验要求 ·· 122
　　二、实验说明与实验拓扑 ·· 122
　　三、实验过程与实验总结 ·· 122
实验 23　STP 的配置 ·· 131
　　一、实验要求 ·· 131
　　二、实验说明 ·· 131
　　三、实验拓扑 ·· 131
　　四、实验过程与实验总结 ·· 131
实验 24　二层交换机链路聚合 ·· 139
　　一、实验要求 ·· 139
　　二、实验说明 ·· 139
　　三、实验拓扑 ·· 139
　　四、配置过程 ·· 139
　　五、实验总结 ·· 140
实验 25　三层交换机链路聚合 ·· 143
　　一、实验要求 ·· 143
　　二、实验说明 ·· 143
　　三、实验拓扑 ·· 143
　　四、配置过程 ·· 143
　　五、实验总结 ·· 145
实验 26　标准 ACL ·· 148
　　一、实验要求 ·· 148
　　二、实验说明 ·· 148
　　三、实验拓扑 ·· 148
　　四、配置过程 ·· 148
　　五、实验总结 ·· 150
实验 27　扩展 ACL ·· 152

  一、实验要求 *152*
  二、实验说明 *152*
  三、实验拓扑 *152*
  四、实验过程及实验总结 *153*

实验 28　基于时间的 ACL *155*
  一、实验要求 *155*
  二、实验说明 *155*
  三、实验拓扑 *155*
  四、实验过程 *155*
  五、实验总结 *158*

实验 29　反射 ACL *159*
  一、实验要求 *159*
  二、实验说明 *159*
  三、实验拓扑 *159*
  四、实验过程 *160*
  五、实验总结 *161*

实验 30　路由器 DHCP 的配置 *164*
  一、实验要求 *164*
  二、实验说明 *164*
  三、实验拓扑 *164*
  四、配置过程 *164*
  五、实验总结 *165*

实验 31　IPv6 静态路由 *167*
  一、实验要求 *167*
  二、实验说明 *167*
  三、实验拓扑 *167*
  四、配置过程 *167*

实验 32　IPv6 RIPng *170*
  一、实验要求 *170*
  二、实验说明 *170*
  三、实验拓扑 *170*
  四、实验过程 *170*
  五、实验总结 *172*

实验 33　IPv6-over-IPv4 隧道 *175*
  一、实验要求 *175*
  二、实验说明 *175*
  三、实验拓扑 *175*
  四、实验过程与实验总结 *175*

实验 34　基于子接口的帧中继配置 *178*

一、实验要求	178
二、实验说明	178
三、实验拓扑	178
四、实验过程	179
五、实验总结	181

实验 35　帧中继的逆向 ARP　182

一、实验要求	182
二、实验说明	182
三、实验拓扑	182
四、配置过程	182
五、实验总结	183

实验 36　RIP Over 帧中继　184

一、实验要求	184
二、实验说明	184
三、实验拓扑	184
四、实验过程	185
五、实验总结	186

实验 37　在多点子接口帧中继下运行 OSPF　191

一、实验要求	191
二、实验说明	191
三、实验拓扑	191
四、实验过程	191
五、实验总结	194

参考文献　196

实验环境说明

由于路由器、交换机、计算机等网络实验设备都非常昂贵，在日常的教学中或个人自学过程中，很多时候使用真实设备完成网络设备配置实验并不方便。为此，现在市面出现了一些网络设备配置的模拟软件，其中有两款软件是学习者们比较爱使用的：一款是 Cisco 的"Cisco Packet Tracer"，另一款是"DynamipsGUI"。

在"实验环境说明"中，主要介绍通过路由器的 console 口（控制台接口）进行连接配置的方法，这是在真实的实验设备环境中使用的；随后再讲述两个模拟软件的使用方法，通过模拟软件，可以随时在单机上非常方便地学习设备的配置方法。

一、用 console 口访问路由器

路由器没有键盘、鼠标和显示屏幕，要配置路由器需要把计算机的串口和路由器的 console 口进行连接。

如图 0-1 所示是 2501 路由器的面板。

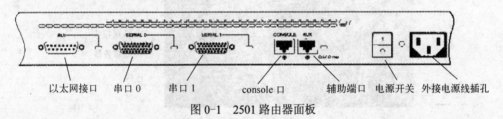

图 0-1　2501 路由器面板

console 口访问路由器的连接方法如图 0-2 所示。

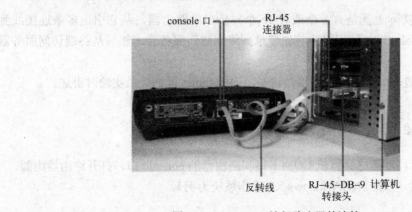

图 0-2　console 口访问路由器的连接

计算机的串口和路由器的 console 口是通过反转线（Kollover Cable）进行连接的，反转线的一端接在路由器的 console 口上，另一端接到一个 DB-9-RJ-45 的转接头上，DB-9 则接

到计算机的串口上，如图 0-2 所示。所谓的反转线就是线两端的 RJ-45 接头上的线序是反的，如图 0-3 所示。

```
Pin 1 -------------- Pin 8
Pin 2 -------------- Pin 7
Pin 3 -------------- Pin 6
Pin 4 -------------- Pin 5
Pin 5 -------------- Pin 4
Pin 6 -------------- Pin 3
Pin 7 -------------- Pin 2
Pin 8 -------------- Pin 1
```

图 0-3　反转线线序

计算机和路由器连接好后，就可以使用各种各样的终端软件配置路由器了。

在实验室里，还可通过终端访问服务器访问路由器，对稍微复杂一点的实验就会用到多台路由器或者交换机，如果通过计算机的串口和它们连接，就需要经常性拔插 console 线，终端访问服务器可以解决这个问题，连接图如图 0-4 所示。

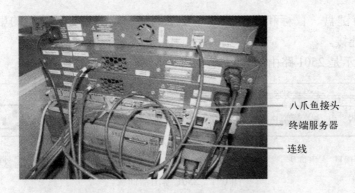

图 0-4　通过终端访问服务器访问路由器

终端访问服务器实际上就是有 8 个或者 16 个异步口的路由器，从它引出多条连接线到各个路由器上的 console 口。使用时，首先登录到终端访问服务器，然后从终端访问服务器再登录到各个路由器。

另外，还可以通过 telnet 访问路由器，这在后面做 telnet 远程登录实验时讲述。

二、通过 console 口连接路由器的操作步骤

① 如图 0-1 所示，连接好计算机 COM 1 口和路由器的 console 口，打开路由器电源。
② 打开超级终端（以经典的 Windows XP 操作系统为例）。

☞说明：

如果在 Windows XP 下没有安装超级终端，或是其他操作系统，如 Windows 7，可以去下载超级终端软件 hyper_terminal-v14.1.0102 并安装使用。

在 Windows 中的【开始】→【程序】→【附件】→【通信】菜单下打开"超级终端"程序，出现如图 0-5 所示窗口。在"名称"文本框中输入名称，例如"Router"，单击【确定】按钮。出现如图 0-6 所示窗口时，在"连接时使用"下拉菜单中选择计算机的 COM 1 口，单击【确定】按钮。

图 0-5 超级终端窗口　　　　　　　　　　图 0-6 选择 COM1 口

③ 在图 0-7 中对 COM1 端口进行如下设置。
- 波特率：9600；
- 数据位：8；
- 奇偶校验：无；
- 停止位：1；
- 数据流控制：无。

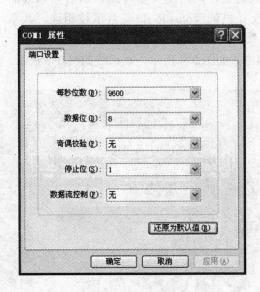

图 0-7 设置 COM1 属性

如果记不住上述参数，最简单的方法是单击下面的"还原为默认值"按钮，即可还原到上述参数。

经过上述两步后，打开路由器即可看见在超级终端上出现路由器的启动过程了，如图 0-8 所示。

图 0-8　启动路由器

路由器启动好后，就可以开始配置路由器了。

三、使用终端访问服务器的方式访问路由器

终端访问服务器的方式访问路由器，一般适用于有网络设备硬件、在实验室中的实验环境。

使用终端访问服务器（就是插有异步模块的路由器，这里用的是 Cisco 2509 路由器作为终端服务器）可以避免在同时配置多台路由器时频繁拔插 console 线。首先，应将需要配置的路由器、交换机的 console 口与八爪线相连，八爪线的另一端与终端服务器相连。

使用八爪线的连接方式如图 0-4 所示，八爪线如图 0-9 所示。

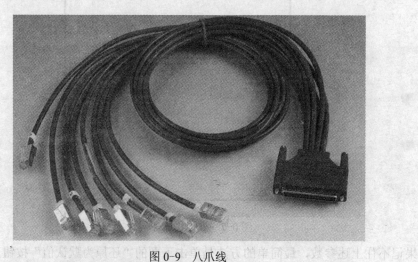

图 0-9　八爪线

下面通过图 0-10 来讲述终端服务器的配置及访问方法。

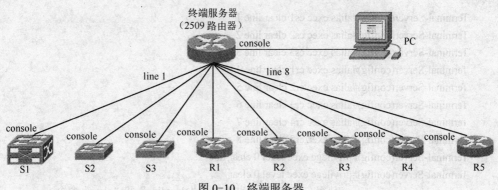

图 0-10 终端服务器

终端服务器的配置过程：

Router(config)#host Terinal-server	//配置路由器名
Terinal-server(config)#enable secret mylab	//配置特权密码，防止他人修改配置
Terinal-server(config)#no ip domain-lookup	//关闭路由器 DNS 查找，以免输入错误
	//命令时长时间等待
Terinal-server(config)#line vty 0 ?	//查看该路由器支持多少 VTY 虚拟终端
<1-15> Last Line number	//支持 15 个
<cr>	
Terinal-server(config)#line vty 0 15	//进入虚拟终端
Terinal-server(config-line)#no login	//允许任何人不需密码就可以 telnet 该终端服务器
Terinal-server(config-line)#logging synchronous	
Terinal-server(config-line)#no exec-timeout	//并且即使长时间不输入命令也不超时退出
Terinal-server(config-line)#exit	
Terinal-server(config)#no ip routing	//关闭路由功能
Terinal-server(config)#line 1 8	//进入线路模式
Terinal-server(config-line)#transport input all	//允许所有传入，包括 telnet 传入
Terinal-server(config-line)#exit	
Terinal-server(config)#int loop0	//进入 Loopback0 接口，并配置 IP 地址
Terinal-server(config-if)#ip add 1.1.1.1 255.255.255.255	
Terinal-server(config-if)#exit	

//以下命令行：从终端服务器控制各路由器，是通过反向 telnet 实现的，此
//时 telnet 的端口号为线路编号加上 2000，例如 line 1，其端口号为 2001，
//如果要控制 line 1 线路上连接的路由器，可以采用"telnet 1.1.1.1 2001"
//命令。然而这样命令很长，为了方便，使用"ip host"命令定义一系列的
//主机名，这样只输入"R1"即可控制 line 1 线路上连接的路由器了

Terinal-server(config)#ip host S1 2001 1.1.1.1
Terinal-server(config)#ip host S2 2002 1.1.1.1
Terinal-server(config)#ip host S3 2003 1.1.1.1
Terinal-server(config)#ip host R1 2004 1.1.1.1
Terinal-server(config)#ip host R2 2005 1.1.1.1
Terinal-server(config)#ip host R3 2006 1.1.1.1
Terinal-server(config)#ip host R4 2007 1.1.1.1
Terinal-server(config)#ip host R5 2008 1.1.1.1

//以下是定义了一系列的命令别名，例如"#alias exec cr1 clear line 4""clear
//line"命令的作用是清除线路

```
Terminal-Server(config)#alias exec cs1 clear line 1
Terminal-Server(config)#alias exec cs2 clear line 2
Terminal-Server(config)#alias exec cs3 clear line 3
Terminal-Server(config)#alias exec cr1 clear line 4
Terminal-Server(config)#alias exec cr2 clear line 5
Terminal-Server(config)#alias exec cr3 clear line 6
Terminal-Server(config)#alias exec cr4 clear line 7
Terminal-Server(config)#alias exec cr5 clear line 8
Terminal-Server(config)#privilege exec level 0 clear line
Terminal-Server(config)#privilege exec level 0 clear
          //使得在用户模式下也能使用"clear line"和"clear"命令
Terminal-Server(config)#exit
Terminal-Server#exit
Terminal-Server>
```

终端访问服务器配置完成后，就可以不用反复地把 console 线来回从一个设备取下，再插到另一个设备的 console 口了，而是在 Terminal-Server>下进行切换。如要去配置路由器 R5，则按以下简单操作即可：

```
Terminal-Server>cr5
Terminal-Server>R5
R5>
```

如果要去配置交换机 S2，则：

```
R5>cs2
R5>S2
S2>
```

以此类推。

四、Cisco Packet Tracer 模拟软件的使用方法

对于很多没有硬件实验环境的学习者，使用模拟软件完成学习是一个非常好的选择。在国内外很多培训机构，甚至学校教学，都在使用模拟软件。

Cisco Packet Tracer 是由Cisco公司发布的一个辅助学习工具，为初学者设计、配置、排除网络故障提供了网络模拟环境。用户可以在软件的图形用户界面上直接使用拖曳方法建立网络拓扑，并可提供数据包在网络中行进的详细处理过程，以观察网络实时运行情况。

当然，此软件也有一些不足之处，部分较复杂的实验不能在 Cisco Packet Tracer 模拟软件上使用（可以使用后面介绍的 **DynamipsGUI** 来完成更复杂的实验）。

1．安装 Cisco Packet Tracer

从 http://www.downxia.com/downinfo/36226.html 下载软件包，并安装好 Cisco Packet Tracer 软件。

2．Cisco Packet Tracer 界面介绍

打开 Cisco Packet Tracer 软件，如图 0-11 所示。

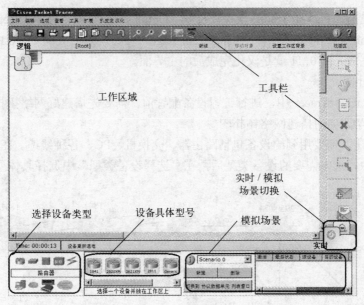

图 0-11 Cisco Packet Tracer 界面

- 工作区域：用于创建网络拓扑，从 4"设备具体型号"中拖入相应设备及线缆后，构建实验拓扑。
- 工具栏：方便快捷完成某些操作，上面横向工具栏与其他软件类似，右边工具栏是本软件常用工具，从上往下，依次是选择、改布局、注释、删除、查看等工具，其中注释、删除、查看几个工具较常用。
- 选择设备类型：可以根据需要选择的设备类型有路由器、交换机、集线器、无线设备、线缆、PC、网云、自定义设备、多用户连接。
- 设备具体型号：在 3"选择设备类型"区域中选择设备类型后，在本区域中可选具体的设备型号。
- 模拟场景：一般不使用。
- 实时/模拟场景切换：实时场景是平时使用最多的一种工作状态，而在模拟场景中，可以通过捕获特定协议以及分析数据包内容、动态观察数据包转发等，一般在正常实验中不使用。

3．软件汉化

对部分英文不太好的使用者来说，英文版的 Cisco Packet Tracer 用起来不太方便，可以对软件进行汉化后再使用。

在 Cisco Packet Tracer 安装包中，找到如图 0-12 所示的汉化包。

打开"packet Tracer5 汉化包"文件夹，如图 0-13 所示，将其中的"chinese.ptl"文件复制到安装目录下的 languages 中，如果在安装 Cisco Packet Tracer 时，是安装在 E 盘的 Program Files\中，则将"chinese.ptl"复制到"E:\Program Files\Cisco Packet Tracer 5.3\languages"下即可。

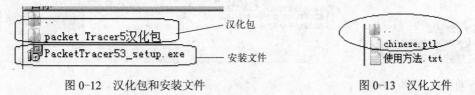

图 0-12 汉化包和安装文件　　　　　　图 0-13 汉化文件

然后运行 Cisco Packet Tracer 软件，打开"Options"对话框，在"Select Language"列表框中选择"chhinese.ptl"，单击"Change Language"按钮，然后关闭 Cisco Packet Tracer 再重启后，汉化完成。图 0-11 就是汉化完成后的主界面。

4. 搭建拓扑图

在 Cisco Packet Tracer 中，进行练习设备配置前，需要先搭建好网络拓扑图，下面根据图 0-14 所示来学习如何搭建网络拓扑图。

在图 0-14 中，需要用到的设备包括路由器、交换机、PC、各种线缆。在图 0-11 的"选择设备类型"中，选中需要的设备类型后，在"选择设备类型"中选择具体型号后，将之拖于工作区域中。

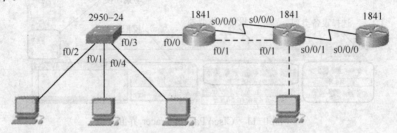

图 0-14　搭建拓扑图

在所有设备放置完毕后，需要将这些设备用线缆连接起来。当选择某种线缆类型后，在需要相连的设备上单击下，弹出该设备所有的接口，如图 0-15 所示是在 PC 上单击后弹出的两个接口。

单击要连接的接口后，线缆即与该设备相连接，然后再去单击另一个设备，同样在弹出的所有接口类型中单击要连接的接口，两设备间的线缆就连接完成。

但是，像图 0-14 中的 1841 路由器，默认只有两个快速以太网接口（Fast Ethernet），而没有串行接口（Serial），要完成两串行口的连接，则需要给路由器添加串口模块。单击图 0-14 中的路由器，弹出如图 0-16 所示的增减模块窗口。

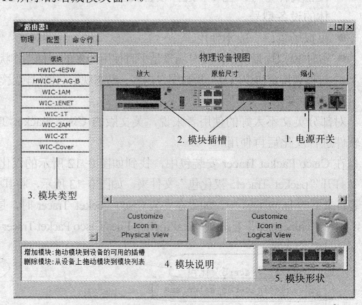

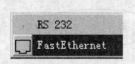

图 0-15　单击 PC 后弹出的接口　　　　　　　　　图 0-16　增减模块窗口

在添加模块之前，应先点一下 1 处的"电源开关"，关闭电源（这类似于拆装真实设备），然后从 3 "模块类型"中选择需要的模块，这里选 WIC-2T，这是一个小口的串行模块，其中的"2"指这个模块中有两个串行接口，将 WIC-2T 拖至 2 处的"模块插槽"后，松开鼠标即完成安装，然后打开电源；拆除模块也同样要关电源，然后将不需要的模块从插槽处拖至 3 "模块类型"处即可。

单击 3 "模块类型"中的任一模块，在 5 "模块形状"处可看见该模块的外观，在 4 "模块说明"处，会显示该模块的一些相关说明信息。

这里需要注意：在关闭电源之前，如果已经对该设备作了部分配置，需先保存，否则在重启后会发现前面的配置已丢失，这和大家日常使用的计算机是一样的。

下面讲述"选项"菜单下的"首选项"，如图 0-17 所示。

图 0-17　首选项

如果按图 0-17 所示搭建拓扑图，则搭建的拓扑如图 0-18 所示，没有设备信息提示。

如图将 0-17 所示几项反选，如图 0-19 所示。

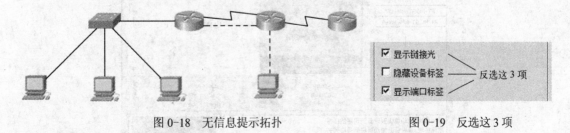

图 0-18　无信息提示拓扑　　　　　　　　图 0-19　反选这 3 项

反选这三项后，则搭建的拓扑如图 0-20 所示，有信息提示。

9

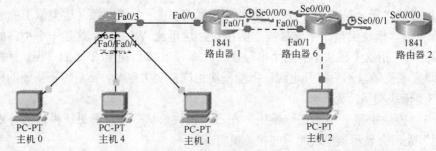

图 0-20　有提示信息的拓扑

这里对比了 3 项常用选项，关于其余选项，大家在练习时可根据需要，勾选和对比这些选项所产生的拓扑图显示效果，可使配置更加方便快捷。

拓扑图搭建完成后，可取名保存起来，以便继续配置时使用，保存拓扑如图 0-21 所示。

图 0-21　保存拓扑

在 Cisco Packet Tracer 中，PC 的配置方法是：单击 PC 图标后，弹出如图 0-22 所示界面。

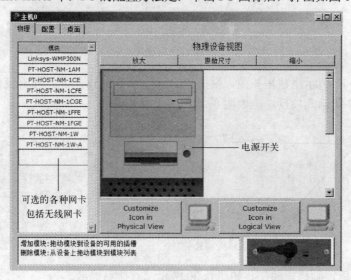

图 0-22　PC 界面

Cisco Packet Tracer 中的 PC 原配有一张快速以太网卡，如需换其他网卡，可在图 0-22 所示界面中，先关电源，然后将滚动条往下拉，出现如图 0-23 所示界面。

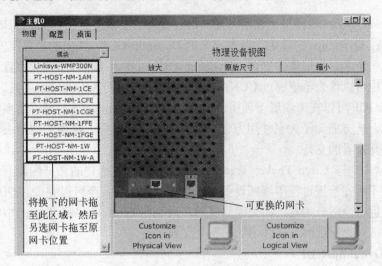

图 0-23　更换网卡

配置 PC 的地址及其他功能，需单击图 0-23 所示的"桌面"选项卡，弹出如图 0-24 所示界面。

图 0-24　桌面选项卡

在桌面选项卡中，可完成各种 PC 上的操作功能，在图 0-24 中标示说明的几项会在后面实验中常用到。具体界面就不在此一一展开演示了。

11

五、DynamipsGUI 的使用方法

Cisco Packet Tracer 软件虽然使用方便，但较复杂的实验不能在 Cisco Packet Tracer 模拟软件上使用。在 Dynamips 的基础上开发出来的 DynamipsGUI，使用的是真实的 Cisco IOS 操作系统构建的一个学习和培训的平台，DynamipsGUI 可作为思科网络实验室管理员的一个补充性的工具，也可以作为希望通过 CCNA/CCNP/CCIE 考试的人们的辅助工具。

DynamipsGUI 可以在未注册下使用，但每次打开 DynamipsGUI 时，需要等到广告弹出的时间。因此，一方面为了大家使用方便，可以去软件标题栏上提供的网址去注册，另一方面也是对软件开发者的支持。

DynamipsGUI 与 Cisco Packet Tracer 相比，其缺点是构建的拓扑图没有 Cisco Packet Tracer 直观，不能在配置过程中增减设备，但其功能强大，甚至可以完成 CCIE 中的大部分实验。当然，也有少部分实验，如无线局域网，只能在 Cisco Packet Tracer 中实现。因此，读者可以根据需要选择这两个软件中的一个来完成实验。

1. 安装 DynamipsGUI

运行 DynamipsGUI 需要的计算机配置越高越好，最低配置如下。

CPU：1.0GHz 以上；

内存：512MB 以上；

空闲硬盘空间：2GB 以上；

操作系统：Windows XP/Windows Server 2003/Windows 7/Linux 等。

DynamipsGUI 的安装过程比较简单，只需注意，在安装完 DynamipsGUI 时，按提示继续安装 WinCap，这是因为 DynamipsGUI 模拟器的底层是需要 WinCap 和 Cisco 的 IOS 来支持。其安装过程在此不再演示。

2. 运行 DynamipsGUI

这里以构建图 0-25 所示的拓扑图为例，来讲述在 DynamipsGUI 中搭建拓扑的方法。

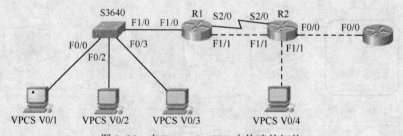

图 0-25 在 DynamipsGUI 中构建的拓扑

打开 DynamipsGUI 后，出现如图 0-26 所示界面。在此界面中，需要根据要构建的拓扑图所需的网络设备的类型和数量，在图 0-26 中进行设置。

在图 0-25 中，有 3 台路由器、1 台交换机和 4 台 PC。因此，在打开 DynamipsGUI 的界面上，应设置路由器个数 3 为交换机个数为 1；在虚拟 PC、7200、3640 前勾选。

其中虚拟 PC 最多可以连接 9 台，可供网络测试使用。在虚拟 PC 前有一个"桥接到 PC"，如果勾选此项，则可使 DynamipsGUI 搭建的路由器与本机相连，使本机作为一台 PC 来使用，其具有比虚拟 PC 更多、更强的功能。

勾选 7200 是确定所使用路由器的类型是 7200 系列的路由器。

此外，DynamipsGUI 这个软件本身没有专门提供交换机模拟，为了完成交换的部分实验，就通过在 3640 路由器上加装交换模块，来实现模拟交换机的功能。因此，勾选 3640 的目的是把它作为交换机使用。

图 0-26　DynamipsGUI 界面

在确定好设备个数和型号后，就需要为设备确定其 IOS 文件的存放位置，以便 DynamipsGUI 为实验需要的设备调用相应 IOS 来使用。如图 0-27 所示，选择下拉菜单中的"7200"。

然后单击"IOS 文件"右边的"浏览"按钮，出现如图 0-28 所示对话框，选择 7200 IOS 文件。

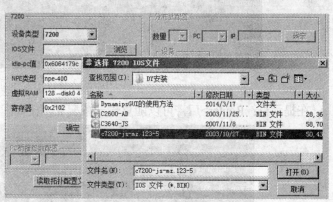

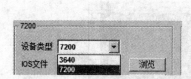

图 0-27　选择"7200"　　　　　　图 0-28　选择 7200 ISO 文件

打开后，此文件包括路径则填入了 IOS 文件右边的方框内，表示已成功选择了 IOS 文件。

接下来，计算 idle-pc 值。在 DynamipsGUI 中有一个默认的 idle-pc 值，但一般不用，如果不重新计算，则会导致 CPU 利用率过高，一般都超过 50%。idle-pc 值计算的方法是：单击 idle-pc 值右边的"计算 idle"按钮，产生如图 0-29 所示窗口。

图 0-29 计算 idle

在按任意键继续之前，先看图 0-29 中的提示："路由器启动后，随意输入点配置，然后按下 ctrl+]+i 即可获取 idle-pc 参数"。现在按下〈Enter〉键，出现图 0-30 所示界面。

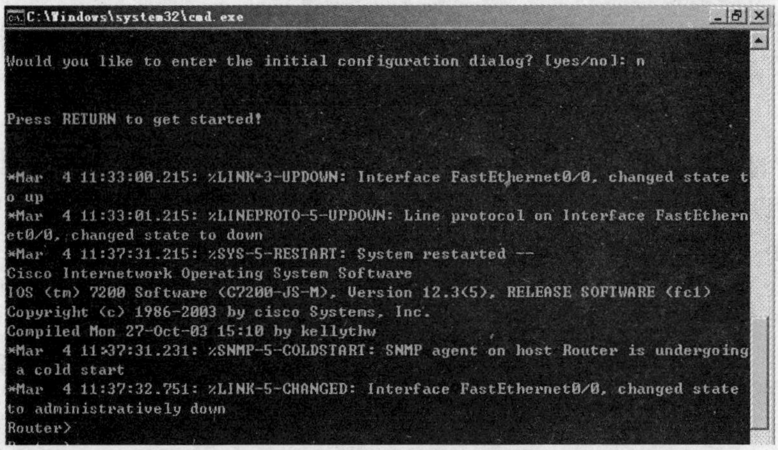

图 0-30 输入任意配置界面

在提示符：Router>后，可以随意输一条配置，比如配置主机名，如图 0-31 所示。

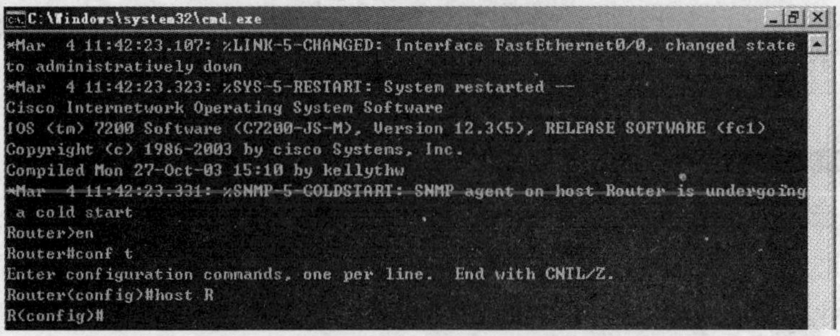

图 0-31 随意输点配置

然后，按前面的提示，按键获取 idle 值，按键方法是：先按下〈Ctrl〉键后不松手，再

14

按下〈]〉键，然后两个键都松手后，再按一下〈i〉键，此时系统提示"Please wait while gathering statistics..."，表示正在采集状态数据，等几秒钟后，出现如图 0-32 所示界面。

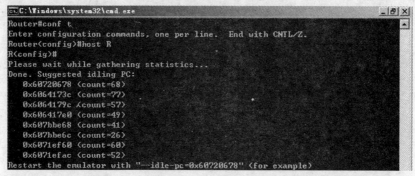

图 0-32　计算 idle 完成

然后，选择 count 值中最大的，这里是 77，把前面的数字全部复制下来，如图 0-33 所示。

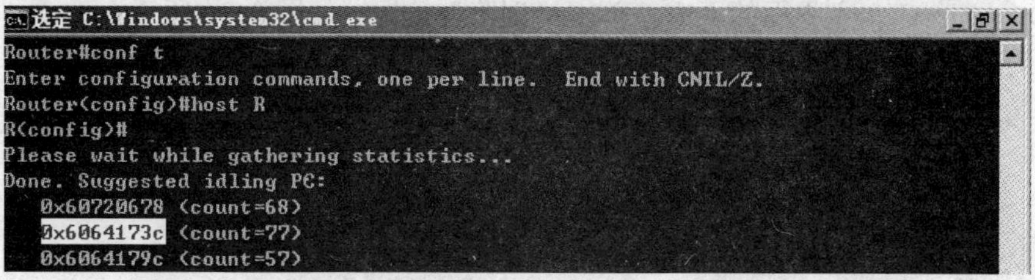

图 0-33　复制最大的 count 对应的数字

复制方法是：单击标题栏上"选定"前的"![]"，在弹出的菜单中单击"编辑"→"标定"，然后选择相应内容，单击右键完成选择，即完成内容的复制（当然，也可以把这个数字直接用手工输入到 idle-pc 值右边的方框中）。

☞说明：

> 在同一台计算机上，对同一 IOS，只需要计算一次 idle 值，以后再做这类实验时，不再需要重新计算 idle 值了。

按同样的方法，再点选如图 0-27 所示的 3640，为交换机选择 IOS 的文件位置，如图 0-34 所示。然后同样需要为 3640 计算 idle-pc 值。计算 idle-pc 值的方法与前面完全相同，在此不重复演示。

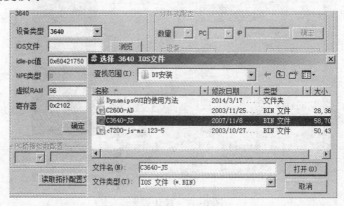

图 0-34　选择 3640 IOS 文件

单击如图0-35所示的"确定"按钮。

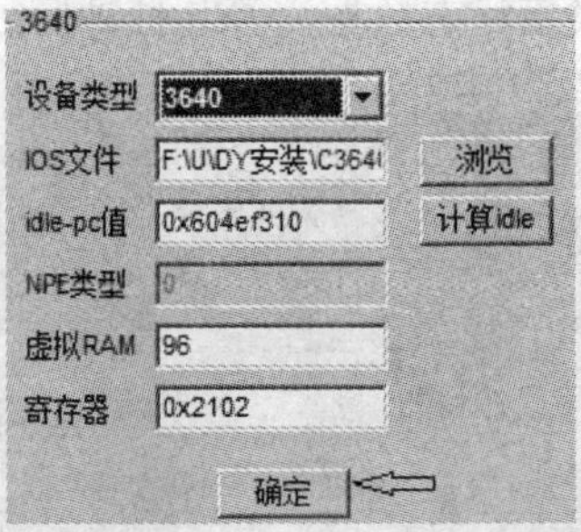

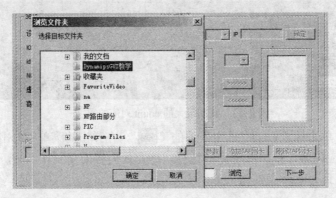

图0-35　单击"确定"按钮

接下来需要为本次搭建的拓扑建一个文件夹，以存放所产生的文件。这里是在E盘上建立了一个"DynamipsGUI 教学"的文件夹（此文件夹可根据自己需要确定位置和名称来建立）。在DynamipsGUI主界面的下方，单击"浏览"按钮，如图0-36所示。

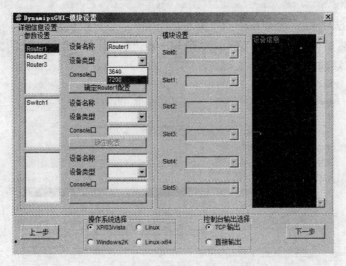

图0-36　确定文件存放位置

然后，单击右下角的"下一步"按钮，进入"模块设置"界面，如图0-37所示。

图0-37　模块设置

在图 0-37 中，选择"Router1"，然后单击"设备类型"下拉菜单，选择"7200"，出现如图 0-38 所示界面。

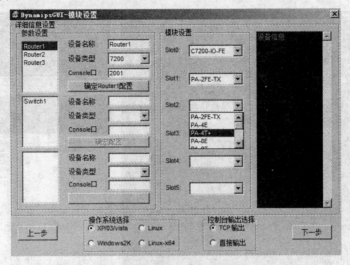

图 0-38　为 Router1 添加模块

根据图 0-25 所示，需要两个快速以太网接口和一个串口。

如图 0-38 所示，先在 Slot0 中选择"C7200-IO-FE"，这里只有唯一一个模块选项；在 Slot1 后选"PA-2FE-TX"，这是添加两个快速以太网模块（PA-4E 是以太网模块）；最后添加串口，选择"PA-4T+"，这模块包括了 4 个串行接口。选择完毕后，单击中间的"确定 Router1 配置"按钮，即可完成 Router1 的模块选择。

按同样的方法，选择 Router2、Router3 的模块。

下一步为 Switch1 选择模块，如图 0-39 所示。

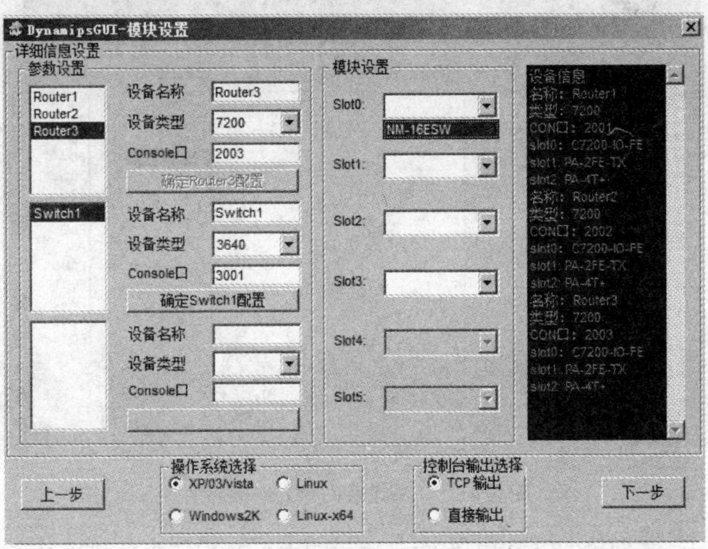

图 0-39　选择交换机模块

这里只需选一个"NM-16ESW"模块就足够了，该模块有 16 个快速以太网接口。

在图 0-39 的下方选择操作系统，这里可根据自己使用的操作系统来选择。勾选"直接输

17

出"项，如果选择 "TCP 输出"，则后面配置路由器时需要使用 telnet 登录，会造成不便，因此建议选择"直接输出"。单击"下一步"按钮，出现如图 0-40 所示的生成文件界面。

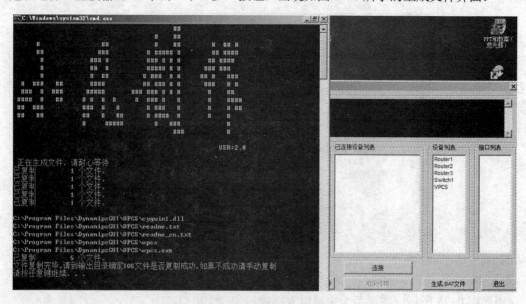

图 0-40　生成文件界面

在图 0-40 中，生成了 5 个文件：3 个路由器、1 个交换机和 1 个虚拟 PC，然后按下〈Enter〉键，生成文件界面消失，出现如图 0-41 所示界面。

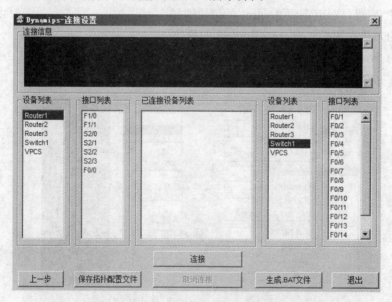

图 0-41　在各设备间建立连接界面

在图 0-41 中，分为"设备列表"和"接口列表"两列，先选择设备，再选需要连接的接口，然后单击中下方的"连接"按钮，即可完成设备间接口的连接。如果连错了，单击下方的"取消连接"即可。连接成功的窗口如图 0-42 所示。

18

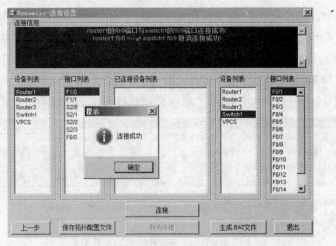

图 0-42 完成两接口间的连接

按同样的方法，完成后面所有接口的连接，完成后如图 0-43 所示。

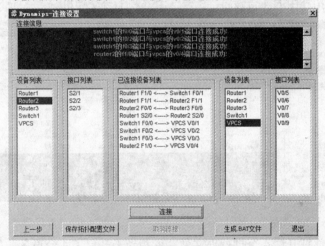

图 0-43 完成所有连接

所有连接完成后，单击右下方的"生成.BAT"文件，如图 0-44 所示。

图 0-44 生成.BAT 文件

在图 0-44 中，确定是否已完成设备间的连接，如果连接完成，单击"确定"按钮。但是，万一在单击了"确定"按钮后，发现还没有连接完，此时就只有重新开始再按刚才的步骤搭建一次了（前面已说明，不需再计算 idle 值了），这也是 DynamipsGUI 不太方便的地方。

最后单击图 0-44 右下角的"退出"按钮，结束拓扑图搭建过程。

在图 0-44 中，单击"确定"按钮之后，会自动打开生成的文件夹，如图 0-45 所示。

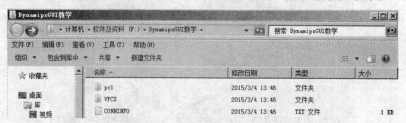

图 0-45 自动打开生成的文件夹

在这个文件夹中，打开 pc1 文件夹，如图 0-46 所示，其中已选定的 4 个批处理文件，对应了刚才搭建的拓扑中的 4 台设备：3 台路由器和 1 台交换机。

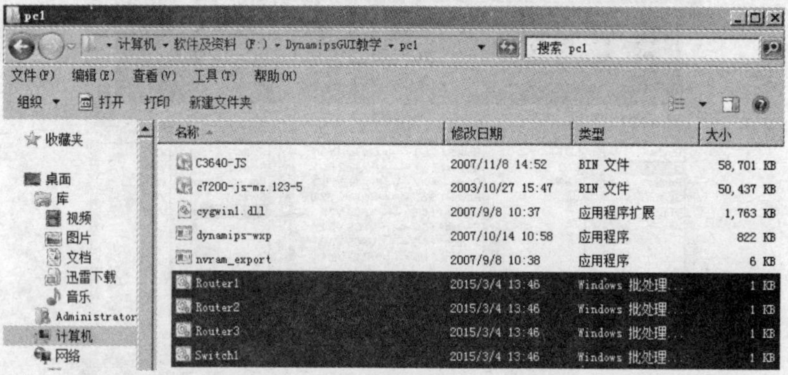

图 0-46 4 台设备

此时，可以全部打开这 4 个批处理文件，就相当于开启了 4 台设备，如图 0-47 所示。

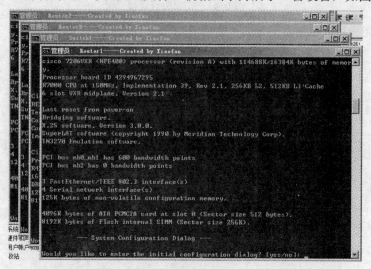

图 0-47 开启 4 台设备

最后一句"Would you like to enter the initial configuration dialog? [yes/no]:",是询问是否进入初始配置对话框,一般回答"n",即进入命令行方式,这是最常用的配置方式,如果回答"y",则进入了初始配置模式,这相当于一个向导模式,系统不停地出现提示问题,不管是否需要,这是一个很让人不舒服的对话,因此,建议在此输入"n",然后进入用户模式,可以正常配置了。

在图0-45中,如果打开"VPCS"文件夹,出现如图0-48所示对话框。

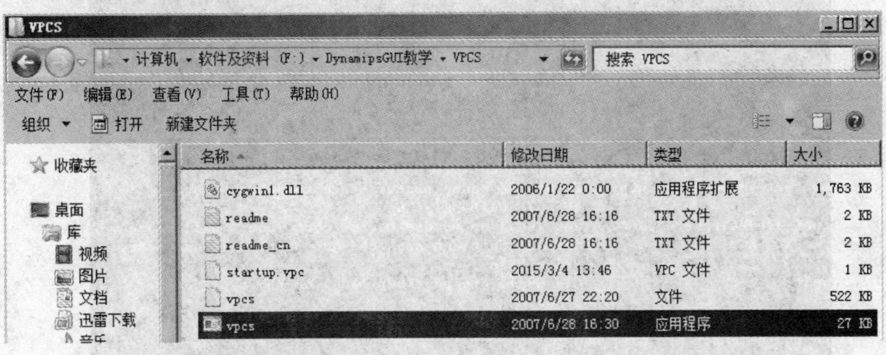

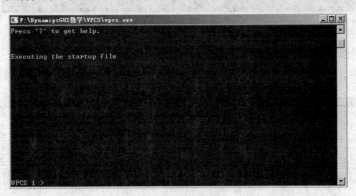

图0-48 "VPCS"文件夹

在"VPCS"文件夹中,如图0-48所示已选定的就是虚拟PC,打开后界面如图0-49所示。

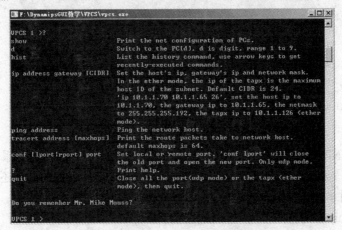

图0-49 虚拟PC

在这里可以输入"?"查看虚拟PC所支持的命令有哪些,如图0-50所示。

图0-50 查看虚拟PC所支持的命令

可见，在虚拟 PC 中可以完成进入其他 PC、输入 IP 地址及网关、ping、tracert 以及配置本地和远程端口等功能。

例如，要进入虚拟的 PC2，则输入"2"，如果给 PC2 配 IP 地址"192.168.2.2"和网关地址"192.168.1.1"，则按如图 0-51 所示输入。

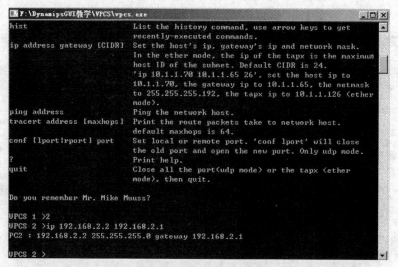

图 0-51 虚拟 PC 的使用

另外，用得比较多的还有 ping 和 tracert 命令，在此就不一一演示了。

在图 0-45 中，还有一个"CONNINFO"记事本文件，打开后，界面如图 0-52 所示。

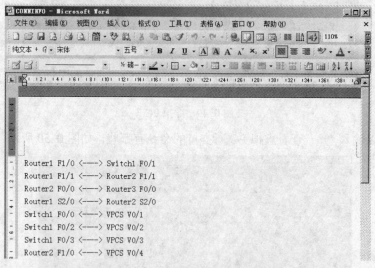

图 0-52 "CONNINFO"文件的内容

可见，"CONNINFO"文件中的内容反映的是各设备间接口的连接关系，因为一旦关闭了 DynamipsGUI 软件后，它不像前面讲的 Cisco Packet Tracer 软件那样还有一个拓扑图存在，所以可以通过本文件中记录的连接关系来查看设备间的接口连接情况。

到此，DynamipsGUI 软件的基本用法介绍完毕。

实验 1　IOS 基本命令

一、实验要求

- 掌握路由器 CLI 的几种模式；
- 掌握路由器 IOS 的基本命令；
- 掌握 CDP 的配置及发现网络拓扑。

二、实验说明

IOS（Internetwork Operating System）是 Cisco 路由器采用的操作系统，用于管理路由器的软硬件资源。对 IOS 的配置有多种方式，如图形界面、Web 页面、管理软件等，但是采用 CLI（Command Line Interface 命令行界面）模式是配置 Cisco IOS 最常用的方法。

本实验讲解了 CLI 的几种常用模式以及路由器 IOS 的常用命令。通过对本实验的学习，可以使读者对配置路由器有一个初步的认识。

三、实验拓扑

本实验采用的拓扑如图 1-1 所示。

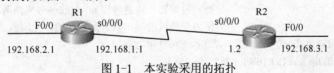

图 1-1　本实验采用的拓扑

四、实验过程

1. 路由器 R1 的配置过程

```
Router>                 //Router 是默认的路由器名，后面的 ">" 表示在用户模式
Router>enable           //从用户模式进入特权模式，输入命令 "enable"
Router#                 //当前处于特权模式，在特权模式下，具有比用户模式更多的命令选项
Router#disable          //输入 "disable" 命令返回用户模式
Router>
Router>?                //在某种模式后，可用 "?" 来查看该模式下支持哪些命令，另外，在命令
                        //的输入过程中，也可以使用 "?" 来查看当前可输入的命令有哪些，下面
                        //的输入表示在用户模式下所支持的命令
Exec commands:
  <1-99>                //Session number to resume
  connect               //Open a terminal connection
  disable               //Turn off privileged commands
```

disconnect	//Disconnect an existing network connection
enable	//Turn on privileged commands
exit	//Exit from the EXEC
logout	// Exit from the EXEC
ping	//Send echo messages
resume	//Resume an active network connection
show	//Show running system information
ssh	//Open a secure shell client connection
telnet	//Open a telnet connection
terminal	//Set terminal line parameters
traceroute	//Trace route to destination

Router>en　　　　　//这里是 enable 命令的简写，从上面查看，在用户模式下的输出可见，以
　　　　　　　　　　//en 开头的命令就只有 enable 一个，在 Cisco IOS 中，只要输入的命令路由
　　　　　　　　　　//器能分辨得出，就可以缩写，养成缩写习惯后，可以将命令简化，使配置
　　　　　　　　　　//更加快捷
Router#conf t　　　　//进入全局配置模式，这是 configure terminal 的简写
Router(config)#host R1　　//这是 hostname R1 的简写
R1(config)#int f0/0　　//进入 f0/0 端口。这是 interface 的简写，在后面的输出中，如果再使用到简
　　　　　　　　　　//写，不再作说明。在学习中，如果有兴趣，可以用键盘上的 Tab 键命令自
　　　　　　　　　　//动补全的方式来查看整个命令的全写。使用 Tab 键自动补全命令的方法：
　　　　　　　　　　//如在输 int f0/0 命令中，在输入 int 后，按 Tab 键，则自动补全为 interface，
　　　　　　　　　　//再输入 f 后，按 Tab 键，自动补全为 interface fastether
R1(config-if)#ip add 192.168.2.1 255.255.255.0　　//给 f0/0 端口配置 IP 地址和子网掩码
R1(config-if)#speed 100　　//配置 f0/0 端口的速率为 100Mbit/s
R1(config-if)#duplex full　　//配置 f0/0 端口为全双工状态
R1(config-if)#no shut　　//激活 f0/0 端口，路由器的端口要能使用，除了前面的一些配置
　　　　　　　　　　//外，很重要一点是要用 no shutdown 来激活，否则该端口不能
　　　　　　　　　　//使用交换机的端口默认是打开的，无须激活即可使用
R1(config-if)#int s0/0/0　　//进入 s0/0/0 串行端口
R1(config-if)#ip add 192.168.1.1 255.255.255.0
R1(config-if)#clock rate ?　　//配时钟频率，查看有哪些时钟频率可配置
Speed (bits per second)
  1200
  2400
  4800
  ……　　　　　　　　//省略部分输出，下同
  2000000
  4000000
<300-4000000>　Choose clockrate from list above
　　　　　　　　　　//从上面列出的时钟频率中选择一个，做实验时可任选一个即可，在实际工程
　　　　　　　　　　//环境中，通过这个数字反映带宽大小，ISP 会显示
R1(config-if)#clock rate 1200
R1(config-if)#no shut
R1(config-if)#end
R1#copy run start　　　　　　//将配置信息存入 NVRAM 中
Destination filename [startup-config]?　　//提示存入的文件名为 startup-config，可直接按〈Enter〉键
Building configuration...

```
[OK]                                         //已保存完毕
R1#show ip interface brief                   //查看各端口 IP 及状态
Interface              IP-Address       OK? Method Status                Protocol

FastEthernet0/0        192.168.2.1      YES manual up                    down

FastEthernet0/1        unassigned       YES unset  administratively down down

Serial0/0/0            192.168.1.1      YES manual down                  down

Serial0/0/1            unassigned       YES unset  administratively down down

Vlan1                  unassigned       YES unset  administratively down down
```

2. 路由器 R2 的配置过程

```
Router>en
Router#conf t
Router(config)#host R2
R2(config)#int f0/0
R2(config-if)#ip add 192.168.3.1 255.255.255.0
R2(config-if)#no shut
R2(config-if)#int s0/0/0
R2(config-if)#ip add 192.168.1.2 255.255.255.0
R2(config-if)#no shut
R2(config-if)#end
R2#ping 192.168.1.1

Type escape sequence to abort.
Sending 5, 100-byte ICMP Echos to 192.168.1.1, timeout is 2 seconds:
!!!!!
Success rate is 100 percent (5/5), round-trip min/avg/max = 3/56/266 ms
//此时，在 R1 和 R2 两端的串行接口都配上 IP 地址及激活端口后，可以 ping 通
```

3. 查看当前正在运行的配置信息

```
R2#show run
Building configuration...

Current configuration : 570 bytes
!
version 12.4
no service timestamps log datetime msec
no service timestamps debug datetime msec
no service password-encryption
!
hostname R2                              //主机名为 R2
!
interface FastEthernet0/0
```

```
   ip address 192.168.3.1 255.255.255.0    //F0/0 端口的 IP 地址
   duplex auto
   speed auto
!
interface FastEthernet0/1
   no ip address
   duplex auto
   speed auto
   shutdown
!
interface Serial0/0/0
   ip address 192.168.1.2 255.255.255.0 //S0/0/0 端口的 IP 地址
!
……
End


R2#wr                         //用 write 命令将配置存入 NVRAM 中
Building configuration...
[OK]
```

4. 查看 NVRAM 中的启动配置文件（即 startup-configure 文件）

```
R2#show start
Using 570 bytes
!
version 12.4
no service timestamps log datetime msec
no service timestamps debug datetime msec
no service password-encryption
!
hostname R2
!
interface FastEthernet0/0
   ip address 192.168.3.1 255.255.255.0
   duplex auto
   speed auto
!
interface FastEthernet0/1
   no ip address
   duplex auto
   speed auto
   shutdown
!
interface Serial0/0/0
   ip address 192.168.1.2 255.255.255.0
!
……
End
```

可见，此时 startup-configure 与 running-configure 是一样的，因为在两次查看之间没有其他输入，如果先存入了 NVRAM 后，再作了其他配置，然后查看 startup-configure 与 running-configure，其内容就有区别了。

5. 查看 s0/0/0 端口状态

R2#show int s0/0/0
Serial0/0/0 is up, line protocol is up (connected)
　　　　//显示物理层已激活，协议已打开，是正常工作状态。此命令可用于故障排查
　　　　//时，查看端口及协议是否正常启用
　Hardware is HD64570
　Internet address is 192.168.1.2/24 　　　//s0/0/0 端口的 IP 地址
　MTU 1500 bytes, BW 1544 kbit, DLY 20000 usec,
　　　　//注意，这里的 BW，表示端口带宽，为 1544kbit；延迟 DLY，为
　　　　//20000μs，这两个参数在后面 eigrp 中会用到
　　reliability 255/255, txload 1/255, rxload 1/255
　Encapsulation HDLC, loopback not set, keepalive set (10 sec)
　　　　// Encapsulation HDLC：封状的协议，Cisco 设备默认是 HDLC，在后
　　　　//面广域网中会讲到
　Last input never, output never, output hang never
　Last clearing of "show interface" counters never
　……

6. 查看所有端口的 IP 地址配置信息

R2#show ip int
FastEthernet0/0 is up, line protocol is down (disabled)
　Internet address is 192.168.3.1/24 　　　//这是 f0/0 的 IP
　Broadcast address is 255.255.255.255
　Address determined by setup command
　MTU is 1500
　……
Serial0/0/0 is up, line protocol is up (connected)
　Internet address is 192.168.1.2/24 　　　//这是 s0/0/0 的 IP
　Broadcast address is 255.255.255.255
　Address determined by setup command
　MTU is 1500
　……

7. 查看系统软硬件状态

R2#show ver
Cisco IOS Software, 1841 Software (C1841-ADVIPSERVICESK9-M), Version 12.4(15)T1, RELEASE SOFTWARE (fc2) 　　　　//IOS 的版本信息
Technical Support: http://www.cisco.com/techsupport
Copyright (c) 1986-2007 by Cisco Systems, Inc.
Compiled Wed 18-Jul-07 04:52 by pt_team

ROM: System Bootstrap, Version 12.3(8r)T8, RELEASE SOFTWARE (fc1)

System returned to ROM by power-on
System image file is "flash:c1841-advipservicesk9-mz.124-15.T1.bin"
　　　　　　　　//当前正在使用的 IOS 文件名

……
A summary of U.S. laws governing Cisco cryptographic products may be found at:
http://www.cisco.com/wwl/export/crypto/tool/stqrg.html
%LINEPROTO-5-UPDOWN: Line protocol on Interface Serial0/0/0, changed state to up
　　　　　　　　//提示 s0/0/0 端口已激活，前面在配置时已手工激活了 f0/0，但这里没提示，是
　　　　　　　　//因为 f0/0 对端没有激活，这是对端没有连其他设备的原因
If you require further assistance please contact us by sending email to
export@cisco.com.

Cisco 1841 (revision 5.0) with 114688K/16384K bytes of memory.
Processor board ID FTX0947Z18E
M860 processor: part number 0, mask 49
2 FastEthernet/IEEE 802.3 interface(s)
2 Low-speed serial(sync/async) network interface(s)
191K bytes of NVRAM.
63488K bytes of ATA CompactFlash (Read/Write)

Configuration register is 0x2102

0x2102 是配置寄存器值，路由器的配置寄存器的值在默认情况下是 0x2102，其中第 4 个字符是用于引导 IOS 时使用。配置寄存器值的第 3 个字符是用于改变路由器加载配置文件的方式，如果配置寄存器值为 0x2102，那么在启动时将会加载保存在 NVRAM 中的配置信息，包括在进入系统时要求输入口令的过程，如果将之改为 0x2142，则启动时将不加载 NVRAM 的内容，也不会要求输入口令，这个一般在口令恢复时临时使用。

　　8. 查看 flash 中存入的信息

　　　　R2#show flash:

　　　　System flash directory:
　　　　File　Length　　Name/status
　　　　　1　　33591768　c1841-advipservicesk9-mz.124-15.T1.bin
　　　　[33847587 bytes used, 30168797 available, 64016384 total]
　　　　　　　　　　//显示 IOS 文件大小、文件名、已占空间、可用空间、总空间
　　　　63488K bytes of processor board System flash (Read/Write)

　　9. 查看 CDP 产生的消息

　　　　R2#show cdp
　　　　Global CDP information:
　　　　　　Sending CDP packets every 60 seconds　　　　//每 60s 发送一次 CDP 消息
　　　　　　Sending a holdtime value of 180 seconds

//设备从相邻设备接收到 CDP 的保留时间为 180s
Sending CDPv2 advertisements is enabled
//当前的 CDP 版本

10. 查看 CDP 邻居信息

R2#show cdp neighbors
Capability Codes: R - Router, T - Trans Bridge, B - Source Route Bridge
                  S - Switch, H - Host, I - IGMP, r - Repeater, P - Phone
Device ID   Local Intrfce   Holdtme   Capability   Platform   Port ID
R1   Ser 0/0/0   132   R   C1841   Ser 0/0/0
              //显示 R2 有一个邻居：R1；R2 通过 s0/0/0 与 R1 相连；收到邻居发送的 CDP
              //消息的有效时间，这是一个倒计时

在 Cisco Packet Tracer 中，部分 CDP 命令不能正常运行，因此，下面的几个命令是在 DynamipsGUI 中完成的，使用的拓扑图如图 1-2 所示。

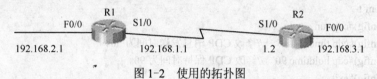

图 1-2　使用的拓扑图

11. 查看 CDP 在各接口的运行情况

R2#show cdp interface
FastEthernet0/0 is up, line protocol is down   //f0/0 端口已激活，协议未工作，因未连对端设备
    Encapsulation ARPA
    Sending CDP packets every 60 seconds   //每 60s 发送一次 CDP 消息
    Holdtime is 180 seconds   //设备从相邻设备接收到 CDP 的保留时间为 180s
Serial1/0 is up, line protocol is up
    Encapsulation HDLC
    Sending CDP packets every 60 seconds
    Holdtime is 180 seconds
Serial1/1 is administratively down, line protocol is down
    Encapsulation HDLC
    Sending CDP packets every 60 seconds
    Holdtime is 180 seconds

12. 关闭 s1/0 端口上的 CDP 协议

R2#conf t
R2(config)#int s1/0
R2(config-if)#no cdp enable   //关闭 s1/0 端口上的 CDP 协议
R2(config-if)#end
R2#show cdp interface   //再次查看 CDP 运行情况，可见，S1/0 端口已经没显示了，表
                                  //示该端口已经不再运行 CDP 协议了
FastEthernet0/0 is up, line protocol is down
    Encapsulation ARPA
    Sending CDP packets every 60 seconds

```
        Holdtime is 180 seconds
    Serial1/1 is administratively down, line protocol is down
        Encapsulation HDLC
        Sending CDP packets every 60 seconds
        Holdtime is 180 seconds
```

13. 在全局模式下关闭 CDP 协议

```
R2#conf t
R2(config)#no cdp run        //在全局模式下关闭 CDP 后，整个设备所有端口都不运行 CDP
R2(config)#exit
R2#show cdp                  //查看 CDP 协议
% CDP is not enabled         //提示 CDP 协议未运行
```

14. 修改 CDP 消息发送时间及保持时间

```
R2#conf t
R2(config)#cdp run
R2(config)#cdp timer 30           //修改 CDP 消息发送时间为 30s
R2(config)#cdp holdtime 90        //修改 CDP 保持时间为 90s
R2(config)#exit
R2#show cdp                       //查看修改后的结果
Global CDP information:
        Sending CDP packets every 30 seconds
        Sending a holdtime value of 90 seconds
        Sending CDPv2 advertisements is    enabled
```

实验 2 路由器的基本操作

一、实验要求

- 设置路由器名：R1。
- 口令设置：

    password 为 cisco1；
    secret 为 cisco2；
    vty 为 cisco3；

要求所有密码都加密。

- 配置以太网口 f0/0 的 IP 为 192.168.1.1/24。
- 设置登录提示信息: welcome to Cisco lab!。
- 对串行口 s0/0/0 进行描述，描述信息为：this is a serial port。
- 将上述信息保存到 NVRAM 中。
- 将实验过程配置写在记事本中进行粘贴。

二、实验说明

本实验主要讲述了关于路由器的常用配置命令，如设置主机名、设置特权口令、设置线路口令、配置端口 IP 以及保存配置等。通过掌握这些命令的使用及功能，为进一步深入学习网络设备配置做准备工作。

三、实验拓扑

本实验使用的拓扑如图 2-1 所示。

图 2-1 路由器的基本操作

四、实验过程

1. 配置

```
Router>en
Router#conf t
Router(config)#host R1              //设置路由器名为 R1
R1(config)#enable password cisco1   //配置特权口令为 cisco1
R1(config)#enable secret cisco2
```

```
                                        //配置特权密码为 cisco2，注意区分大小写。当 password 和
                                        //secret 都配置时，secret 有效，因为它安全性更高
R1(config)#line vty 0 4                 //进入虚拟线路，对 0~4 号线路进行配置
R1(config-line)#password cisco3         //设置虚拟线路口令为 cisco3
R1(config-line)#login                   //要求登录时验证口令
R1(config-line)#exit
R1(config)#service password-encryption  //对所有口令加密，提高安全性
R1(config)#int f0/0
R1(config-if)#ip add 192.168.1.1 255.255.255.0
                                        //配置 f0/0 端口的 IP 地址和子网掩码
R1(config-if)#no shut
R1(config-if)#exit
R1(config)#banner motd &                //配置登录提示信息，以"&"开头
welcome to Cisco lab!                   //提示信息的内容
&                                       //提示信息以"&"结束

R1(config)#int s0/0/0
R1(config-if)#description this is a serial port
                    //对 s0/0/0 端口进行描述，描述信息是 this is a serial port
R1(config-if)#end
R1#wr
                    //保存到 NVRAM 中，此命令与"copy run start"功能相同，只是使用"copy
                    //run start"可以另起文件名，而"wr"则使用默认文件名
R1#
```

2. 查看配置（删除了其中的部分空行输出）

```
R1#show run
Building configuration...

Current configuration : 736 bytes
!
version 12.4
no service timestamps log datetime msec
no service timestamps debug datetime msec
service password-encryption
!
hostname R1
!
enable secret 5 $1$mERr$yG9qv7LLYVv0YzwRYtdTM/
                    //这是用 secret 加密的密码，在查看配置时显示为乱码，因此安全性更高
enable password 7 0822455D0A1654
                    //这是用 password 加密的口令，如果没有再使用"service password-encryption"
                    //对口令加密，则以明文显示
!
interface FastEthernet0/0
 ip address 192.168.1.1 255.255.255.0
```

```
   duplex auto
   speed auto
!
interface FastEthernet0/1
   no ip address
   duplex auto
   speed auto
   shutdown
!
interface Serial0/0/0
   description this is a serial port        //端口描述信息
   no ip address
   shutdown
!
interface Serial0/0/1
   no ip address
   shutdown
!
interface Vlan1
   no ip address
   shutdown
!
ip classless
!
banner motd ^C
welcome to Cisco lab!                       //登录提示信息
^C
!
line con 0
line vty 0 4
   password 7 0822455D0A1656
                    //如果没有使用"service password-encryption"对线路口令再加密，
                    //则以明文显示线路口令
   login
!
end
```

实验 3　将真实 PC 与 DynamipsGUI 桥接

一、实验要求

- 掌握在 Windows 7 中添加环回网卡的方法；
- 掌握给环回网卡添加 IP 地址的方法；
- 掌握在 DynamipsGUI 中将 PC 桥接到路由器的方法。

二、实验说明

在 DynamipsGUI 中有两种方式与 PC 相连，一种是虚拟 PC，另一种是桥接到 PC，如图 3-1 所示。

图 3-1　两种连接 PC 的方式

其中虚拟 PC 的功能远不如真实 PC 强大，不能满足某些实验的需求。而桥接到 PC 则是将使用 DynamipsGUI 软件的计算机与 DynamipsGUI 实现连接，从而提供真实计算机的功能，例如将在实验 4 中讲述的端口镜像和协议分析，就必须用到真实 PC 与 DynamipsGUI 才能实现。

三、操作过程

本实验主要以操作过程截图的方式，讲述 PC 与 DynamipsGUI 桥接过程。

在本机上添加环回网卡，如图 3-2 所示。

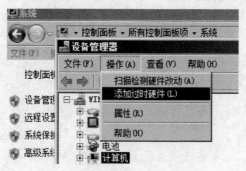

图 3-2　添加环回网卡

选择设备管理器的"操作"→"添加过时硬件"选项，然后出现如图 3-3 所示界面。

在图 3-3 中单击"下一步"按钮，进入图 3-4 所示界面。

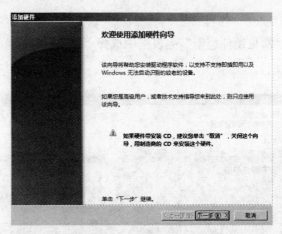

图 3-3　添加硬件向导

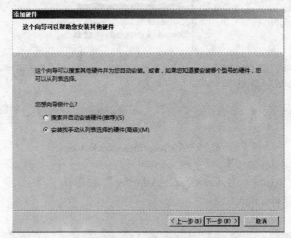

图 3-4　勾选"安装我手动从列表选择的硬件（高级）(m)"选项

勾选"安装我手动从列表选择的硬件（高级）(m)"选项后，单击"下一步"按钮进入如图 3-5 所示界面。

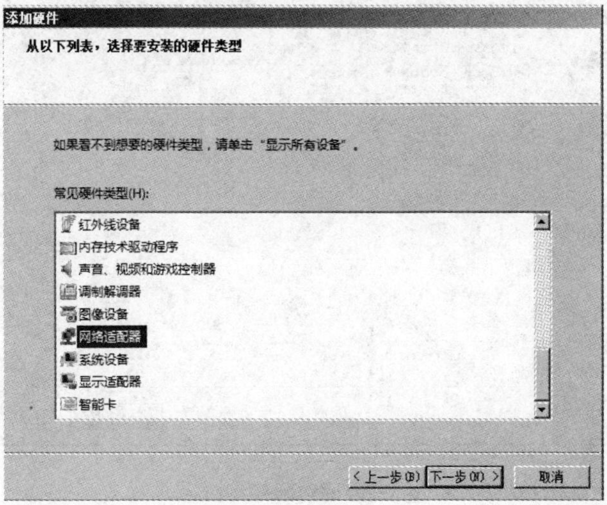

图 3-5　添加硬件

在图 3-5 中,在"常见硬件类型"列表框中选择"网络适配器"后进入如图 3-6 所示界面。

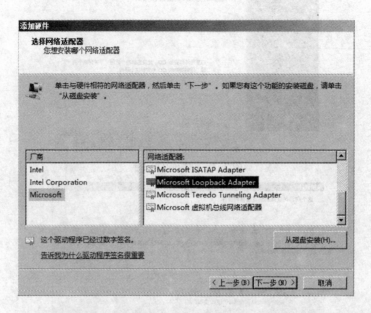

图 3-6　选择"Microsoft Loopback Adapter"

在图 3-6 中选择"Microsoft Loopback Adapter"后单击"下一步"按钮,进入图 3-7 和图 3-8 所示的安装硬件界面。

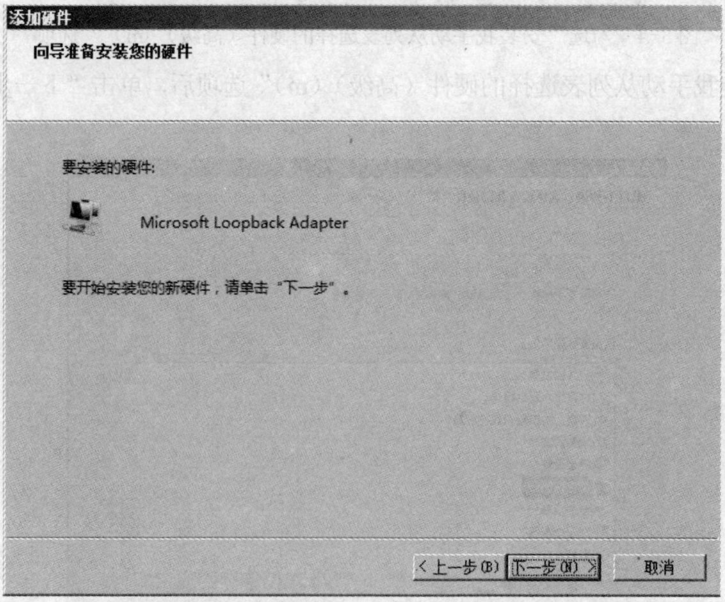

图 3-7　安装硬件(1)

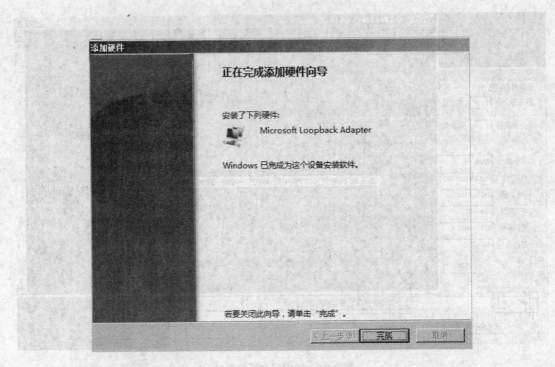

图 3-8　安装硬件（2）

在图 3-8 中，单击"完成"按钮后，在设备管理器中会看到已安装好的环回网卡，如图 3-9 所示。

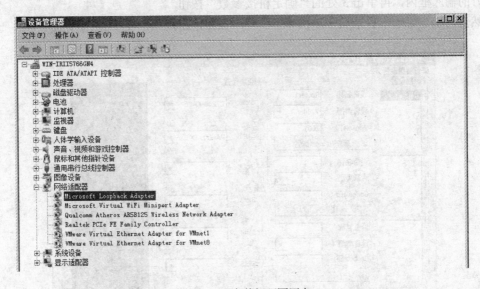

图 3-9　已安装好环回网卡

然后，打开 DynamipsGUI 软件，首先按图 3-1 所示选择"桥接到 PC"，其余的如选择路由器等设备、确定 IOS、计算 Idle-pc 值的方法，均在实验 1 中讲述过。

如图 3-10 所示确定网卡桥接参数。

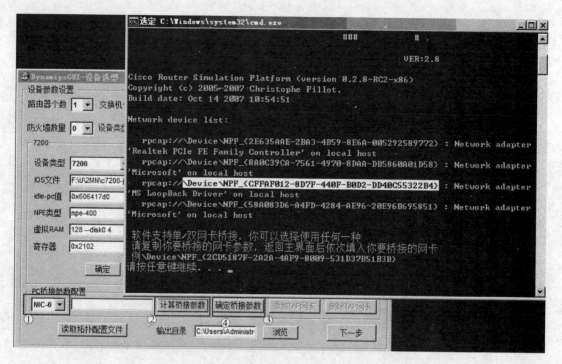

图 3-10 确定网卡桥接参数

在图 3-10 中，先在 1 处选择"NIC-0"网卡，然后单击 2 处"计算桥接参数"按钮，弹出图 3-10 右边所示窗口，复制当前所选定的桥接参数（对应的是 MS LoopBack Driver）到 2 处上方的输入框内，再单击 3 处的"确定桥接参数"按钮。

然后，确定好输出目录后，单击"下一步"按钮，出现如图 3-11 所示界面。

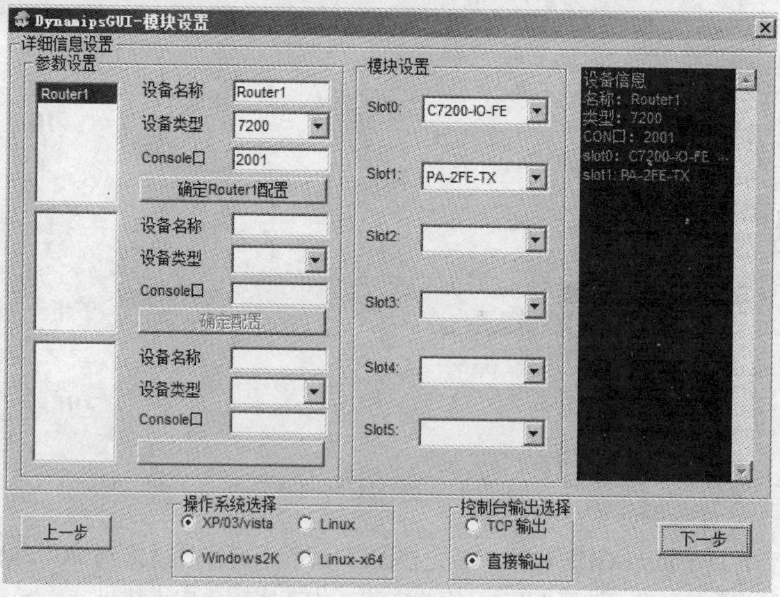

图 3-11 确定网络设备模块

按图3-11所示确定好模块后,单击"下一步"按钮,进入图3-12所示界面。

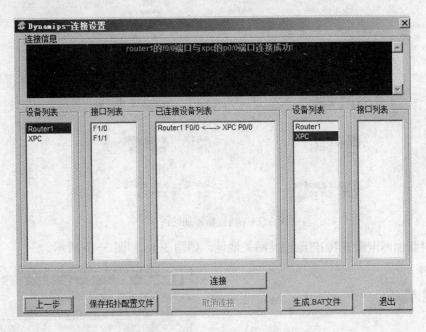

图3-12 连接路由器与所桥接的PC

按图3-12所示连接路由器与所桥接的PC后,单击"生成.BAT文件"按钮。打开所生成的.BAT文件,出现如图3-13所示界面。

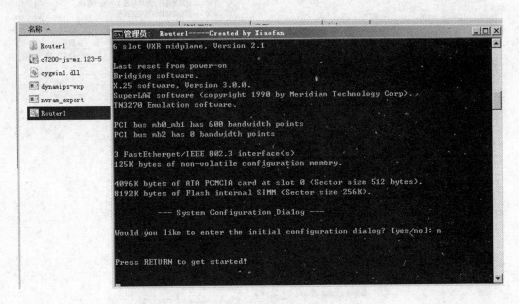

图3-13 启动路由器

按图3-13所示,启动路由器
在本机上的网络属性中,找到新添加的网卡,这里是"本地连接5",如图3-14所示。

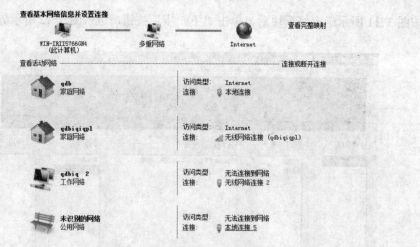

图 3-14　找到新添加的网卡

然后对新加网卡配置其 IP 地址及网关地址，如图 3-15 和图 3-16 所示。

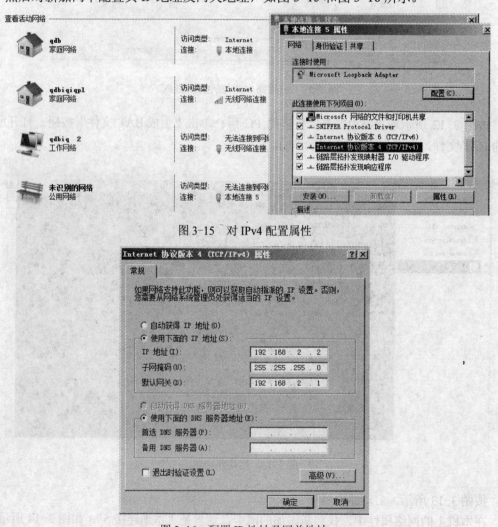

图 3-15　对 IPv4 配置属性

图 3-16　配置 IP 地址及网关地址

然后，在路由器上，配置与桥接 PC 相连的接口 IP，然后测试与 PC 的连通性，如图 3-17 所示。

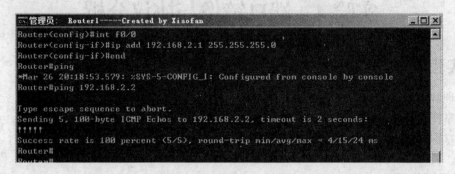

图 3-17 测试与 PC 的连通性

可见，本机与 DynamipsGUI 软件中的路由器已成功桥接。

完成本操作后，在本机上可以完成比虚拟 PC 更多的功能，例如可以通过端口镜像，并借助协议分析仪，来对需要分析的流经某端口的流量进行查看分析。

实验 4 端口镜像与协议分析

一、实验要求

- 理解端口镜像功能；
- 掌握端口镜像的配置方法；
- 掌握 Sniffer 网络协议分析仪的基本使用方法。

二、实验说明

端口镜像（Port Mirroring）是在交换机上，通过将一个或多个源端口的数据流量转发到某一个指定端口来实现对网络的监听镜像的功能，以对被监控流量进行故障定位、流量分析、流量备份等。镜像目标端口（监控端口）一般直接与监控主机（安装有 Sniffer 等分析软件）相连。

端口镜像监视到进出网络的所有数据包，供安装了监控软件的管理服务器抓取数据，如网吧需提供此功能把数据发往公安部门审查。而企业出于信息安全、保护公司机密的需要，也需要网络中有一个端口能提供这种实时监控功能。在企业中用端口镜像功能，可以很好地对企业内部的网络数据进行监控管理，在网络出现故障的时候，可以很好地做到故障定位。

本实验通过讲解端口镜像的配置方法，并利用 telnet 远程登录使用 TCP 的三次握手机制，来进一步讲述了 Sniffer 网络协议分析仪的基本使用方法。

通过本实验的学习，使读者在后面的学习中能更深层次地分析网络数据的传递过程。

三、实验拓扑

实验拓扑如图 4-1 所示。

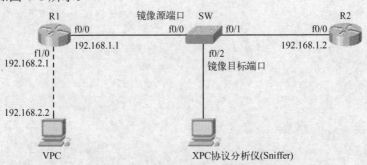

图 4-1 端口镜像与协议分析实验拓扑

四、实验过程

1. R1 的配置过程

```
Router>en
```

```
Router#conf t
Router(config)#host R1
R1(config)#int f0/0
R1(config-if)#ip add 192.168.1.1 255.255.255.0
R1(config-if)#no shut
R1(config)#int f1/0
R1(config-if)#ip add 192.168.2.1 255.255.255.0
R1(config-if)#no shut
R1(config)#router rip                    //配置 RIP 路由协议,关于 RIP 将在后面的实验中讲述
R1(config-router)#netw 192.168.2.0
R1(config-router)#netw 192.168.1.0
R1(config-router)#end
```

2. R2 的配置过程

```
Router>en
Router#conf t
Router(config)#host R2
R2(config)#int f0/0
R2(config-if)#ip add 192.168.1.2 255.255.255.0
R2(config-if)#no shut
R2(config-if)#exit
R2(config)#enable password ccnp
R2(config)#line vty 0 4
R2(config-line)#password ccna
R2(config-line)#login
R2(config-line)#exit
R2(config)#router rip
R2(config-router)#netw 192.168.1.0
R2(config-router)#
```

3. SW 的配置过程

```
Router>en
Router#conf t
Router(config)#host SW
SW(config)#no cdp run
            //关闭 CDP（Cisco Discovery Protocol，思科发现协议），以免在配置中被不断产
            //生的 CDP 信息干扰
SW(config)#monitor session 1 source interface fastEthernet 0/0 ?
            //配置端口镜像源端口,即被监控端口。这里可查看相关参数
    ,    Specify another range of interfaces        //可同时对多个端口镜像
    -    Specify a range of interfaces              //可同时对某个范围内端口镜像
    -    both   Monitor received and transmitted traffic  //镜像进入和外出两方向的流量
    rx   Monitor received traffic only              //只镜像进入方向的流量
    tx   Monitor transmitted traffic only           //只镜像外出方向的流量
    <cr>
SW(config)#monitor session 1 source interface fastEthernet 0/0 both
            //配置为镜像进入和外出两个方向流量
```

SW(config)#monitor session 1 destination interface fastEthernet 0/2
　　　　//配置端口镜像目标端口，即监控口
SW(config)#

4．在 VPC（虚拟 PC）上的配置

VPCS 1 >ip 192.168.2.2 192.168.2.1
　　//配虚拟 PC 的 IP 地址和网关地址

5．在 XPC（本机）上的设置

在本机上的设置如图 4-2 所示。

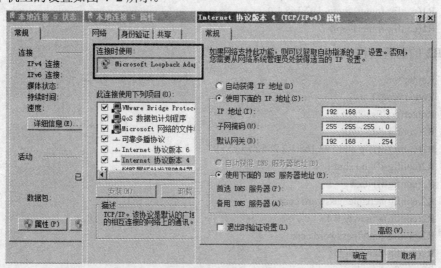

图 4-2　本机上的设置

在图 4-2 中，是对环回网卡配置 IP 地址，一般可配置在与 R1、R2 相连接口的同一网段，但是，也可以配置在其他网段，甚至让其自动获取 IP 地址也可以实现数据的捕获。

6．在本机上启动 Sniffer

启动 Sniffer 后，出现选择与 Sniffer 绑定的网卡的窗口，如图 4-3 所示。

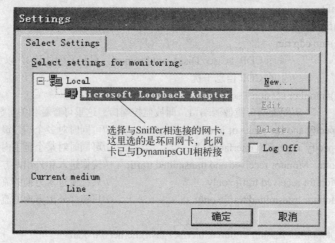

图 4-3　选择与 Sniffer 绑定的网卡

在图 4-3 中，选择"Microsoft Loopback Adapter"环回网卡（第一次启动 Sniffer 时，一般还会有本机的物理网卡），然后单击"确定"按钮，即进入图 4-4 所示界面。

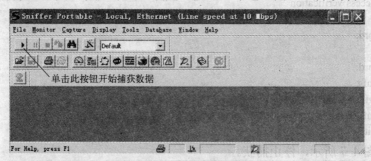

图 4-4　开始捕获数据

如果 Sniffer 在安装过程中设置不当，可能单击"确定"按钮后不能正常进入图 4-4 所示界面，这时可以按图 4-5 所示的方式操作。

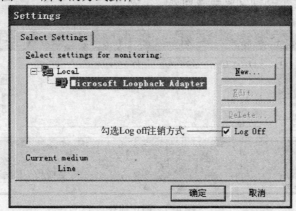

图 4-5　勾选 Log Off

勾选"Log Off"选项后，单击"确定"按钮，可以进入图 4-6 所示界面，然后单击"File"文件菜单下的"Log On"命令登录。

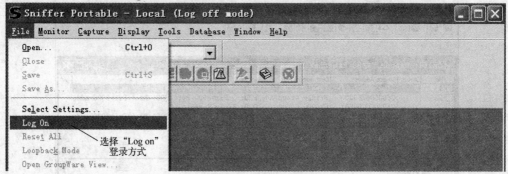

图 4-6　选择 Log On

在图 4-6 中，单击"Log On"后也可进入图 4-4 所示界面（正常情况下不需要图 4-5 和图 4-6 所示的操作过程）。

此时，我们到路由器 R1 上对 R2 进行远程登录。

　　R1#telnet 192.168.1.2

Trying 192.168.1.2 ... Open

User Access Verification

Password:
R2>en
Password:
R2#conf t
R2(config)#host RR
RR(config)#exit
RR#exit

然后，按图 4-7 所示，停止捕获。

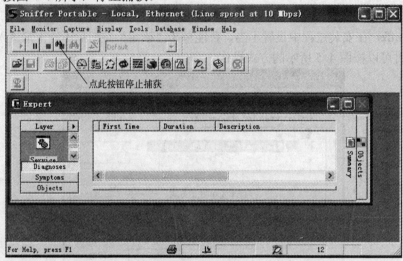

图 4-7　停止捕获

在停止捕获后，出现图 4-8 所示界面，单击"Decode"按钮。

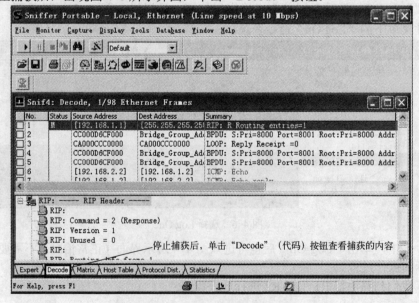

图 4-8　单击"Decode"按钮

在单击"Decode"按钮后,可以对捕获的代码进行查看分析。在此先看一下 telnet 建立 TCP 连接时"三次握手"的过程示意图,如图 4-9 所示。

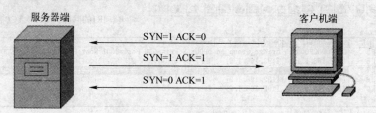

图 4-9 TCP 的"三次握手"

在图 4-9 中,提供的是 TCP 建立时的三次握手过程中 SYN 和 ACK 的参数值,这是大多数参考资料都可以看到的,在此可以通过 Sniffer 来分析和证实这些参数值,以及 TCP 的三次握手的具体过程。

TCP 的三次握手过程如图 4-10～图 4-12 所示。

在图 4-12 中第三次握手没有注释,可参照前两次握手自行分析。

除了图 4-10～图 4-12 中所标识的内容外,还可以从 Sniffer 中看到更多关于网络协议的相关信息,例如在 TCP 段中,包括源端口、目标端口、序列号等信息,通过 Sniffer 协议分析,可以充分证实很多资料以文字方式所述的理论知识。

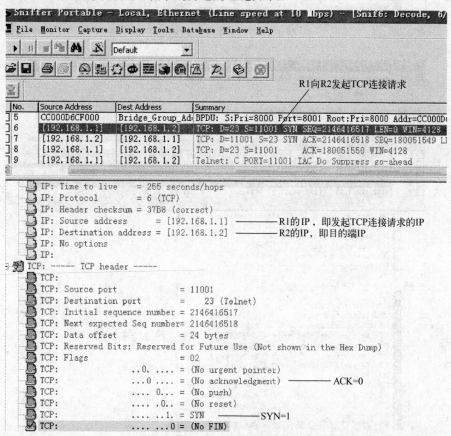

图 4-10 TCP 的第一次握手

图 4-11 TCP 的第二次握手

图 4-12 TCP 的第三次握手

然后，继续查看远程登录后面的操作，如图 4-13 和图 4-14 所示。

图 4-13 远程登录过程一

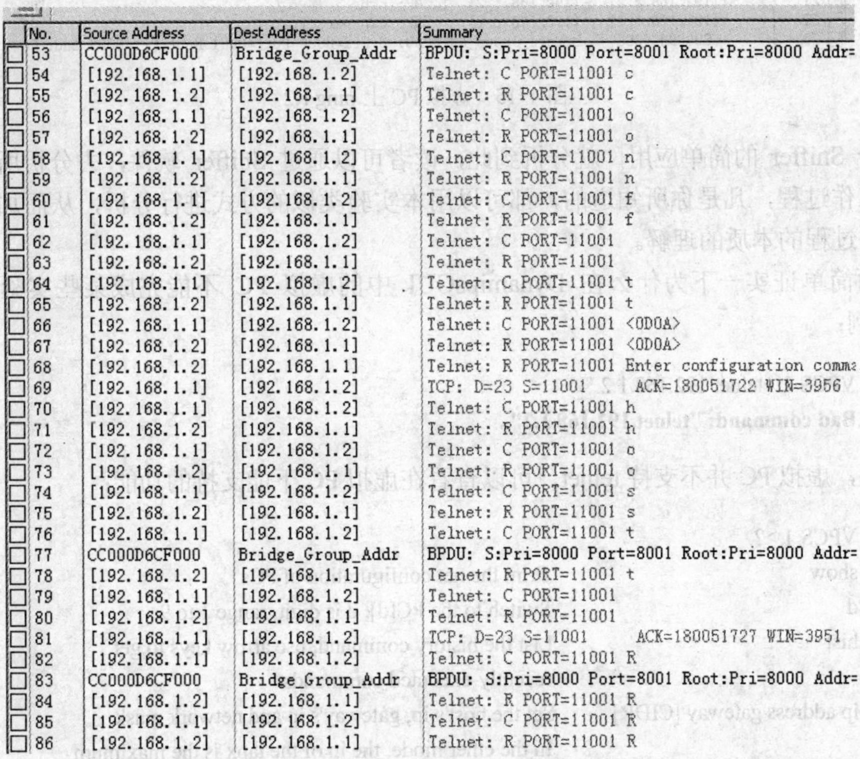

图 4-14 远程登录过程二

在图 4-14 中显示的是在 R1 上输入的命令和 R1 应答所收到的命令字符序列。

在虚拟 PC 上 ping R2：

```
VPCS 1 >ping 192.168.1.2
192.168.1.2 icmp_seq=1 time=78.000 ms
192.168.1.2 icmp_seq=2 time=62.000 ms
192.168.1.2 icmp_seq=3 time=63.000 ms
192.168.1.2 icmp_seq=4 time=78.000 ms
192.168.1.2 icmp_seq=5 time=62.000 ms
```

同样，镜像到本机后，通过 Sniffer 分析其 ping 的过程，如图 4-15 所示。

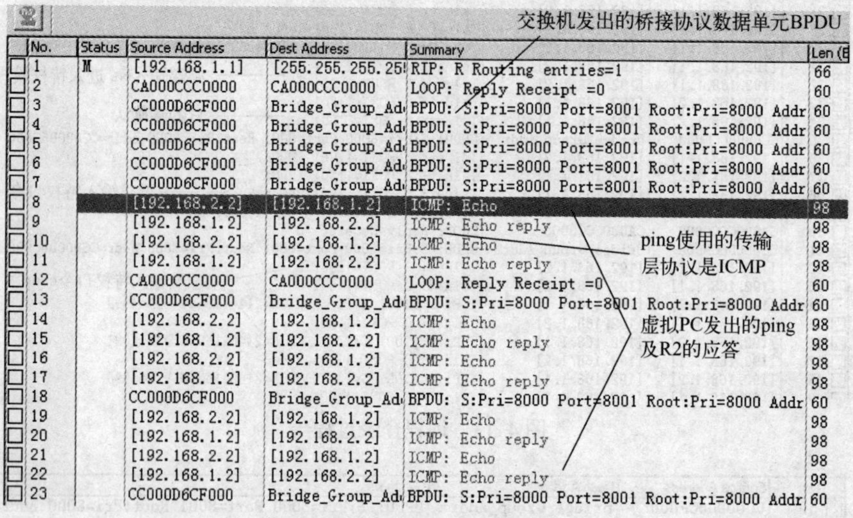

No.	Status	Source Address	Dest Address	Summary	Len (B
1	M	[192.168.1.1]	[255.255.255.25	RIP: R Routing entries=1	66
2		CA000CCC0000	CA000CCC0000	LOOP: Reply Receipt =0	60
3		CC000D6CF000	Bridge_Group_Ad	BPDU: S:Pri=8000 Port=8001 Root:Pri=8000 Addr	60
4		CC000D6CF000	Bridge_Group_Ad	BPDU: S:Pri=8000 Port=8001 Root:Pri=8000 Addr	60
5		CC000D6CF000	Bridge_Group_Ad	BPDU: S:Pri=8000 Port=8001 Root:Pri=8000 Addr	60
6		CC000D6CF000	Bridge_Group_Ad	BPDU: S:Pri=8000 Port=8001 Root:Pri=8000 Addr	60
7		CC000D6CF000	Bridge_Group_Ad	BPDU: S:Pri=8000 Port=8001 Root:Pri=8000 Addr	60
8		[192.168.2.2]	[192.168.1.2]	ICMP: Echo	98
9		[192.168.1.2]	[192.168.2.2]	ICMP: Echo reply	98
10		[192.168.2.2]	[192.168.1.2]	ICMP: Echo	98
11		[192.168.1.2]	[192.168.2.2]	ICMP: Echo reply	98
12		CA000CCC0000	CA000CCC0000	LOOP: Reply Receipt =0	60
13		CC000D6CF000	Bridge_Group_Ad	BPDU: S:Pri=8000 Port=8001 Root:Pri=8000 Addr	60
14		[192.168.2.2]	[192.168.1.2]	ICMP: Echo	98
15		[192.168.1.2]	[192.168.2.2]	ICMP: Echo reply	98
16		[192.168.2.2]	[192.168.1.2]	ICMP: Echo	98
17		[192.168.1.2]	[192.168.2.2]	ICMP: Echo reply	98
18		CC000D6CF000	Bridge_Group_Ad	BPDU: S:Pri=8000 Port=8001 Root:Pri=8000 Addr	60
19		[192.168.2.2]	[192.168.1.2]	ICMP: Echo	98
20		[192.168.1.2]	[192.168.2.2]	ICMP: Echo reply	98
21		[192.168.2.2]	[192.168.1.2]	ICMP: Echo	98
22		[192.168.1.2]	[192.168.2.2]	ICMP: Echo reply	98
23		CC000D6CF000	Bridge_Group_Ad	BPDU: S:Pri=8000 Port=8001 Root:Pri=8000 Addr	60

图 4-15 虚拟 PC 上 ping R2

关于 Sniffer 的简单应用，就分析到此，读者可以通过 Sniffer 软件，去分析所有的网络协议的工作过程，凡是你所知道的，都可以用本实验类似的方式进行分析，从而加深对网络协议工作过程的本质的理解。

下面简单证实一下为什么在 DynamipsGUI 中的虚拟 PC 不能完成某些实验，这里以 telnet 为例：

VPCS 1 >telnet 192.168.1.2

Bad command: "telnet 192.168.1.2"。

可见，虚拟 PC 并不支持 telnet，可以查看在虚拟 PC 下能支持的功能：

VPCS 1 >?

show	Print the net configuration of PCs.
d	Switch to the PC[d], d is digit, range 1 to 9.
hist	List the history command, use arrow keys to get recently-executed commands.
ip address gateway [CIDR]	Set the host's ip, gateway's ip and network mask. In the ether mode, the ip of the tapx is the maximum host ID of the subnet. Default CIDR is 24.

	'ip 10.1.1.70 10.1.1.65 26', set the host ip to 10.1.1.70, the gateway ip to 10.1.1.65, the netmask to 255.255.255.192, the tapx ip to 10.1.1.126 (ether mode).
ping address	Ping the network host.
tracert address [maxhops]	Print the route packets take to network host. default maxhops is 64.
conf [lport|rport] port	Set local or remote port. 'conf lport' will close the old port and open the new port. Only udp mode.
?	Print help.
quit	Close all the port(udp mode) or the tapx (ether mode), then quit.

Do you remember Mr. Mike Muuss?
VPCS 1 >

可见虚拟 PC 的功能非常有限，可通过将 PC 桥接到 DynamipsGUI 上以提供更多、更强大的功能。

清空交换机的端口镜像：

SW(config)#no monitor session 1

实验 5 静态路由的配置

一、实验要求

- 掌握静态路由的配置方法；
- 理解路由表的含义；
- 了解下一跳地址静态路由和送出接口静态路由的区别；
- 掌握扩展 ping 的使用。

二、实验说明

在小型网络中，网络结构基本处于稳定状态，为了减少路由器 CPU、RAM 的资源占用、减少对网络带宽的占用，增加网络的可控性和安全性，可以采用静态路由。

静态路由也有其缺点，即配置与维护较麻烦，配置时易出错，对于大型网络和网络拓扑结构变化大的网络一般不适用静态路由；同时，配置静态路由也需要网络管理员对整个网络的情况完全了解。

可以通过在实验中对比，每增加一个网络，需要静态路由条目就是多条，在学习静态路由配置时，最好要像本实验那样，采用三个以上的路由器，才能更好地理解和熟练掌握静态路由的配置。

三、实验拓扑

本实验使用的拓扑图如图 5-1 所示。

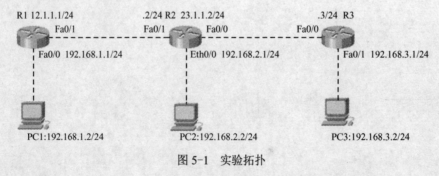

图 5-1 实验拓扑

四、配置过程

1. 路由器 R1 的配置过程

```
Router>en
Router#conf t
Router(config)#host R1
```

```
R1(config)#int f0/1
R1(config-if)#ip add 12.1.1.1 255.255.255.0        //给 f0/1 端口配 IP 地址
R1(config-if)#no shut
R1(config-if)#int f0/0
R1(config-if)#ip add 192.168.1.1 255.255.255.0
R1(config-if)#no shut
R1(config-if)#exit
R1(config)#ip route 192.168.2.0 255.255.255.0 12.1.1.2
        //配置到达 192.168.2.0/24 网络的静态路由，通过 IP 为 12.1.1.2 的端口转发
```

静态路由在全局模式下配置，命令格式为：

```
R1(config)#ip route [destination_network] [mask] [next-hop_address/ exit-interface] [administrative_distance] [peimanent]
```

其中的参数说明如下。

- ip route：配置静态路由固定命令。
- destination_network：静态路由要到达的目标网络。
- mask：目标网络的掩码。
- next-hop_address：转发数据包的下一跳路由器的入口 IP 地址，此时管理距离为 1。
- exit-interface：转发数据包的本路由器出口名称，此时管理距离为 0。
- administrative_distance：指定静态路由管理距离。
- peimanent：参数包括 reject 和 blackhole。reject 指明为不可达路由；blackhole 指明为黑洞路由。如果没有指明 reject 或 blackhole，则默认为可达路由。

```
R1(config)#ip route 192.168.3.0 255.255.255.0 12.1.1.2
        //配置到达 192.168.3.0/24 网络的静态路由，通过 IP 为 12.1.1.2 的端口转发，管理距离为 1
R1(config)#ip route 23.1.1.0 255.255.255.0 f0/1
        //配置到达 23.1.1.0/24 网络的静态路由，通过本路由器外出端口 f0/1 转发，管理距离为 0
R1(config)#exit
R1#wr        //保存配置
Building configuration...
[OK]
R1#
```

2. 路由器 R2 的配置过程

```
Router>en
Router#conf t
Router(config)#host R2
R2(config)#int f0/1
R2(config-if)#ip add 12.1.1.2 255.255.255.0
R2(config-if)#no shut
R2(config-if)#int f0/0
R2(config-if)#ip add 23.1.1.2 255.255.255.0
R2(config-if)#no shut
R2(config-if)#int e0/0/0
R2(config-if)#ip add 192.168.2.1 255.255.255.0
```

```
R2(config-if)#no shut
R2(config-if)#exit
R2(config)#ip route 192.168.1.0 255.255.255.0 12.1.1.1
R2(config)#ip route 192.168.3.0 255.255.255.0 f0/0
R2(config)#exit
R2#wr
Building configuration...
[OK]
R2#
```

3. 路由器 R3 的配置过程

```
Router>en
Router#conf t
Router(config)#host R3
R3(config)#int f0/0
R3(config-if)#ip add 23.1.1.3 255.255.255.0
R3(config-if)#no shut
R3(config-if)#int f0/1
R3(config-if)#ip add 192.168.3.1 255.255.255.0
R3(config-if)#no shut
R3(config-if)#exit
R3(config)#ip route 192.168.1.0 255.255.255.0 f0/0
R3(config)#ip route 192.168.2.0 255.255.255.0 f0/0
R3(config)#ip route 12.1.1.0 255.255.255.0 23.1.1.2
R3(config)#exit
R3#wr
Building configuration...
[OK]
R3#
```

4. 查看路由表

```
R3#show ip route
Codes: C - connected, S - static, I - IGRP, R - RIP, M - mobile, B - BGP
       D - EIGRP, EX - EIGRP external, O - OSPF, IA - OSPF inter area
       N1 - OSPF NSSA external type 1, N2 - OSPF NSSA external type 2
       E1 - OSPF external type 1, E2 - OSPF external type 2, E - EGP
       i - IS-IS, L1 - IS-IS level-1, L2 - IS-IS level-2, ia - IS-IS inter area
       * - candidate default, U - per-user static route, o - ODR
       P - periodic downloaded static route
              //以上是路由代码部分
Gateway of last resort is not set

     12.0.0.0/24 is subnetted, 1 subnets
S       12.1.1.0 [1/0] via 23.1.1.2
     23.0.0.0/24 is subnetted, 1 subnets
C       23.1.1.0 is directly connected, FastEthernet0/0
S       192.168.1.0/24 is directly connected, FastEthernet0/0
```

  S 192.168.2.0/24 is directly connected, FastEthernet0/0
  C 192.168.3.0/24 is directly connected, FastEthernet0/1

  这里一共有五条路由。其中，有代码"C"的路由条目表示直连路由，即与本路由器直接相连的网络。例如，"C 23.1.1.0 is directly connected, FastEthernet0/0"表示 23.1.1.0 网络通过 FastEthernet0/0 端口直接与路由器 R3 相连。

  有代码"S"的表通过静态配置的路由。这里有两种类型的静态路由：一类如"192.168.1.0/24 is **directly connected**, FastEthernet0/0"，注意其中的 **directly connecte** 表示带送出接口的静态路由，要到达目标网络 192.168.1.0/24，其送出接口是 FastEthernet0/0；另一类如"S 12.1.1.0 **[1/0]** via 23.1.1.2"，注意其中的"**[1/0]**"，1 表示管理距离为 1，0 是度量值（静态路由和直连路由的度量值都是规定的，为 0），要到达目标网络 12.1.1.0，是通过 IP 地址为 23.1.1.2 的端口转发。

5. 在 PC1 的配置过程

  在主机 PC1 的"桌面"选项卡中，单击"IP 配置"按钮，弹出如图 5-2 所示界面。

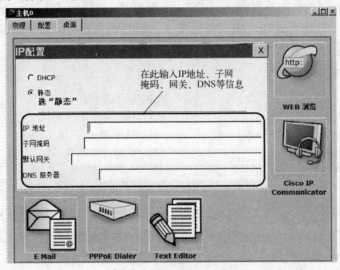

图 5-2 IP 配置相关信息

  结合拓扑图 5-1 所示 IP 地址，按图 5-2 中的说明给 PC1 配置好相应的 IP 地址，如图 5-3 所示。

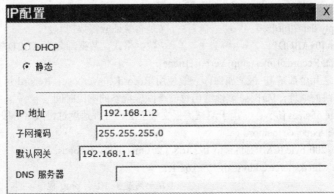

图 5-3 给 PC1 配置 IP 地址

按同样方法，配置 PC2 和 PC3 的 IP 地址信息。

6. Ping

Ping 有两种形式：标准 Ping 和扩展 Ping，下面以从 R1 Ping PC3 为例，来分别演示这两类 Ping 命令。

（1）标准 Ping

  R1#ping 192.168.3.2   //Ping 后带目标 IP 地址

  Type escape sequence to abort.
  Sending 5, 100-byte ICMP Echos to 192.168.3.2, timeout is 2 seconds:
  ...!!
  Success rate is 40 percent (2/5), round-trip min/avg/max = 6/8/11 ms

  R1#

从 R1 Ping PC3 时，可以发现"...!!"，前面有"..."表示还在进行 ARP 地址解析，最后两个"!!"表示已解析成功，能正常通信了。以后再去进行 Ping 测试时，就可直接通信，不用再解析了。

（2）扩展 Ping

  R1#ping   //ping 后不带参数
  Protocol [ip]:   //选择协议，默认 IP 协议
  Target IP address: 192.168.3.2 //输入目标 IP
  Repeat count [5]: 10   //发出 ping 包的次数
  Datagram size [100]: 200
    //设置 ping 包的大小，如果怀疑报文由于延迟过长或者分段失败而丢失，则可以提高
    //报文的大小
  Timeout in seconds [2]:
    //超时时间，如果怀疑超时是由于响应过慢而不是报文丢失，则可以提高该值
  Extended commands [n]: y   //是否进一步使用扩展命令
  Source address or interface: 192.168.1.1 //输入源 IP
  Type of service [0]:   //服务类型
  Set DF bit in IP header? [no]:
    //是否在 IP 头部设置 DF 位，通过设置 DF 位禁止分段，即使是报文超过了路由器定义
    //的 MTU 也禁止分段
  Validate reply data? [no]:   //是否验证应答数据
  Data pattern [0xABCD]:   //数据格式，数据部分为 ABCD 模式
  Loose, Strict, Record, Timestamp, Verbose[none]:
    //这几个都是 IP 报文属性，一般使用 Record 和 Verbose。Record 可以用来记录报文每一
    //跳的地址，Verbose 属性给出每一个回应应答的响应时间
  Sweep range of sizes [n]: //用于测试大报文被丢失、处理速度过慢或者分段失败等故障
  Type escape sequence to abort.
  Sending 10, 200-byte ICMP Echos to 192.168.3.2, timeout is 2 seconds:
  Packet sent with a source address of 192.168.1.1
  .!!!!!!!!!   //一共发了 10 次 ping 包
  Success rate is 90 percent (9/10), round-trip min/avg/max = 6/10/16 ms

实验 6　浮动静态路由

一、实验要求

- 掌握浮动静态路由的配置方法；
- 掌握备份静态路由工作原理。

二、实验说明

浮动静态路由是指给常用的通信路由以静态路由的方式另指定一条备份路由。当常用通信的路由出现故障时，则选用作为备份的静态路由。

例如在本实验中，路由器 R1 和 R2 间的优选路由是通过两 F0/0 相连的以太网链路，而通过两 S1/0 相连的串行链路配置为浮动静态路由，当以太网链路发生故障时，选用串行链路（浮动静态路由），在以太网链路恢复后，则优选以太网链路。

三、实验拓扑

本实验使用的拓扑图如图 6-1 所示。

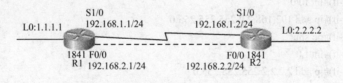

图 6-1　实验拓扑

四、实验配置

1. 在 R1 上的配置过程

```
Router>
Router>en
Router#conf t
Router(config)#host R1
R1(config)#int s1/0
R1(config-if)#ip add 192.168.1.1 255.255.255.0
R1(config-if)#no shut
```

```
R1(config-if)#int f0/0
R1(config-if)#ip add 192.168.2.1 255.255.255.0
R1(config-if)#no shut
R1(config-if)#int L0
R1(config-if)#ip add 1.1.1.1 255.255.255.0
R1(config-if)#exit
R1(config)#ip route 2.2.2.0 255.255.255.0 192.168.1.2 200
        //将基于串行接口的静态路由管理距离设为 200，默认为 1
R1(config)#ip route 2.2.2.0 255.255.255.0 192.168.2.2
        //再配置一条基于以太网接口的静态路由，不修改管理距离
R1(config)#end
R1#
```

这里配置了两条静态路由，通过配置不同的管理距离，使路由器在选择路由时，优选管理距离最小的一条，上述配置过程中，基于以太网接口的静态路由为优选路由，其管理距离默认为 1，当它出现故障时再选择基于串行接口的静态路由。在工作中，串行链路虽然速度慢，但能实现基本通信，不彻底影响工作。此时管理员将以太网链路恢复后，就可恢复优选以太网链路。

2. 在 R2 上的配置过程

```
Router>
Router>en
Router#conf t
Router(config)#host R2
R2(config)#int s1/0
R2(config-if)#ip add 192.168.1.2 255.255.255.0
R2(config-if)#no shut
R2(config-if)#int f0/0
R2(config-if)#ip add 192.168.2.2 255.255.255.0
R2(config-if)#no shut
R2(config-if)#int L0
R2(config-if)#ip add 2.2.2.2 255.255.255.0
R2(config-if)#exit
R2(config)#ip route 1.1.1.0 255.255.255.0 s1/0 200
R2(config)#ip route 1.1.1.0 255.255.255.0 f0/0
R2(config)#end
```

查看 R2 的路由表：

```
R2#show ip route
……                                          //此处省略代码输出部分，后同
     1.0.0.0/24 is subnetted, 1 subnets
S       1.1.1.0 is directly connected, FastEthernet0/0   //可见路由走的是 f0/0 端口
     2.0.0.0/24 is subnetted, 1 subnets
C       2.2.2.0 is directly connected, Loopback0
C    192.168.1.0/24 is directly connected, Serial1/0
C    192.168.2.0/24 is directly connected, FastEthernet
```

查看 R1 的路由表：

R1#show ip route
……
  1.0.0.0/24 is subnetted, 1 subnets
C   1.1.1.0 is directly connected, Loopback0
  2.0.0.0/24 is subnetted, 1 subnets
S   2.2.2.0 [1/0] via 192.168.2.2   //路由走的 192.168.2.2，为以太网端口 IP
C 192.168.1.0/24 is directly connected, Serial1/0
C 192.168.2.0/24 is directly connected, FastEthernet0/0

把 R1 的 f0/0 端口关闭，查看路由的变化情况：

R1#conf t
R1(config)#int f0/0
R1(config-if)#shut   //关闭了 f0/0 端口，制造"故障"

再去查看 R1 的路由表：

R1#show ip route
……
  1.0.0.0/24 is subnetted, 1 subnets
C   1.1.1.0 is directly connected, Loopback0
  2.0.0.0/24 is subnetted, 1 subnets
S   2.2.2.0 [100/0] via **192.168.1.2** //路由选择的 192.168.1.2 是串行接口，浮动路由被
                //放入路由表
C 192.168.1.0/24 is directly connected, Serial1/0

再重新打开 f0/0，相当于故障修复：

R1(config)#int f0/0
R1(config-if)#no shut
R1(config-if)#end
R1#show ip route
……
  1.0.0.0/24 is subnetted, 1 subnets
C   1.1.1.0 is directly connected, Loopback0
  2.0.0.0/24 is subnetted, 1 subnets
S   2.2.2.0 [1/0] via **192.168.2.2**   //可见，此时路由优选的是以太网接口，浮动路由又
                //恢复为备份路由了
C 192.168.1.0/24 is directly connected, Serial1/0
C 192.168.2.0/24 is directly connected, FastEthernet0/0

实验 7 默认路由

一、实验要求

- 掌握默认静态路由的配置方法；
- 掌握默认静态路由工作原理。

二、实验说明

某企业内部有很多人都需要上 Internet，并且所访问的网站大多不同，那么要到达的目标网络就会千变万化，管理员不可能在企业中的路由器上去通过手动来为不同的到达网络一一配置静态路由，当然，也不能使用动态路由将企业内部网络向公众网络公开。在这种情况下，就可以使用默认路由。

默认路由是指路由器在路由表中找不到目的网络时，让路由器通过默认路由来转发数据包。

三、实验拓扑

本实验使用的拓扑图如图 7-1 所示。

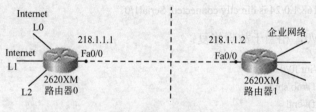

图 7-1　实验拓扑

如图 7-1 所示，路由器 0 表示 ISP 端路由器，路由器 1 表示企业边界路由器，使用 f0/0 端口相连接。

四、实验配置

1. 路由器 0 的基本配置

```
Router>en
Router#conf t
Router(config)#host ISP
ISP(config)#int f0/0
ISP(config-if)#ip add 218.1.1.1 255.255.255.0
ISP(config-if)#no shut
```

再配置 3 个虚拟接口，表示 Internet 中的目标网络：

```
ISP(config-if)#int L0
ISP(config-if)#ip add 1.1.1.1 255.255.255.0
ISP(config-if)#int L1
ISP(config-if)#ip add 2.2.2.2 255.255.255.0
ISP(config-if)#int L2
ISP(config-if)#ip add 3.3.3.3 255.255.255.0
```

2．路由器 1 的基本配置

```
Router>en
Router#conf t
Router(config)#host R1
R1(config)#int f0/0
R1(config-if)#ip add 218.1.1.2 255.255.255.0
R1(config-if)#no shut
```

完成基本配置后，根据实验 6 所述，企业边界路由器 1 可以 ping 通 ISP 端路由器 0 的 f0/0 端口，但不能 ping 通路由器 0 的 L0、L1、L2 端口，说明路由器 1 不能与 Internet 连通。

查看路由器 1 的路由表：

```
R1#show ip route
……
C    218.1.1.0/24 is directly connected, FastEthernet0/0
```

在企业边界路由器 1 上配置一条默认路由：

```
R1(config)#ip route 0.0.0.0 0.0.0.0 218.1.1.1
    //与静态路由类似，这里的外出端口可以是与 ISP 路由器相连的外出端口，也可以是 ISP 路由
    //器端口。其中作为网络的 0.0.0.0 表示任意网络，相对应的子网掩码为 0.0.0.0。
```

然后查看路由表：

```
R1#show ip route
……
C    218.1.1.0/24 is directly connected, FastEthernet0/0
S*   0.0.0.0/0 [1/0] via 218.1.1.1
```

可见路由器 1 新增了一条到达 0.0.0.0 网络的路由，前面标记是"S*"，说明这是默认路由。然后测试到路由器 0 的 L0 端口的连通性：

```
R1#ping 1.1.1.1
Type escape sequence to abort.
Sending 5, 100-byte ICMP Echos to 1.1.1.1, timeout is 2 seconds:
!!!!!
```

同样，路由器 1 也能 ping 通 L1 和 L2 端口。

可见，有了默认路由，当企业的边界路由器上收到内网上任何需要到达在路由表中查不到的目的网络，都将被送往默认的端口，从而由 ISP 的路由器转发进入 Internet。

实验 8 RIPv1

一、实验要求

- 掌握 RIP 的配置方法；
- 理解 RIP 路由表含义；
- 掌握被动接口的作用及配置方法。

二、实验说明

RIP（Routing Information Protocols，路由信息协议）分为两个版本 RIPv1 和 RIPv2。通过本实验，可以学习到 RIPv1 的基本配置方法和动态路由表各项的含义，了解传递子网路由的条件、RIPv1 更新路由的方式、等价路由和被动接口等相关知识。

三、实验拓扑

本实验使用的拓扑图如图 8-1 所示。

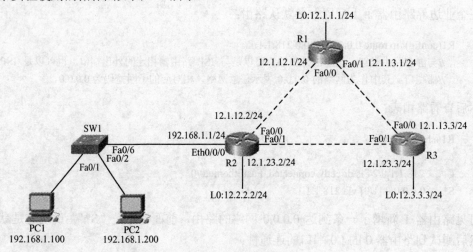

图 8-1 RIPv1 实验拓扑

四、实验配置

1. 路由器 1 的基本配置过程

```
Router>en
Router#conf t
Router(config)#host R1
```

```
R1(config)#int f0/0
R1(config-if)#ip add 12.1.12.1 255.255.255.0
R1(config-if)#no shut
R1(config-if)#int f0/1
R1(config-if)#ip add 12.1.13.1 255.255.255.0
R1(config-if)#no shut
R1(config)#int L0
R1(config-if)#ip add 1.1.1.1 255.255.255.0
R1(config-if)#exit
R1(config)#router rip                    //配置 RIP 协议
R1(config-router)#ver 1                  //指定 RIP 版本为 1
R1(config-router)#netw 12.0.0.0          //公告网络，在公告 RIPv1 网络时，只公告主网
R1(config-router)#end
```

2．路由器 2 的基本配置过程

```
Router>en
Router#conf t
Router(config)#host R1
R2(config)#int f0/0
R2(config-if)#ip add 12.1.12.2 255.255.255.0
R2 (config-if)#no shut
R2 (config-if)#int f0/1
R2 (config-if)#ip add 12.1.23.2 255.255.255.0
R2 (config-if)#no shut
R2(config-if)#int e0/0/0
R2(config-if)#ip add 192.168.1.1 255.255.255.0
R2(config-if)#no shut
R2(config)#int L0
R2(config-if)#ip add 12.2.2.2 255.255.255.0
R2(config-if)#exit
R2(config)#router rip
R2(config-router)#netw 192.168.1.0
R2(config-router)#netw 12.0.0.0
R2(config-router)#end
```

3．路由器 3 的基本配置过程

```
Router>en
Router#conf t
Router(config)#host R3
R3(config)#int f0/0
R3(config-if)#ip add 12.1.13.3 255.255.255.0
R3(config-if)#no shut
R3(config-if)#int f0/1
R3(config-if)#ip add 12.1.23.3 255.255.255.0
R3(config-if)#no shut
R3(config)#int L0
```

R3(config-if)#ip add 12.3.3.3 255.255.255.0
R3(config-if)#exit
R3(config)#router rip
R3(config-router)#netw 12.0.0.0
R3(config-router)#end

4. 查看路由器 R1 的路由表

R1#show ip route
……
12.0.0.0/24 is subnetted, 6 subnets
C 12.1.1.0 is directly connected, Loopback0
C 12.1.12.0 is directly connected, FastEthernet0/0
C 12.1.13.0 is directly connected, FastEthernet0/1
R 12.1.23.0 [120/1] via 12.1.12.2, 00:00:00, FastEthernet0/0
     [120/1] via 12.1.13.3, 00:00:23, FastEthernet0/1
R 12.2.2.0 [120/1] via 12.1.12.2, 00:00:00, FastEthernet0/0
R 12.3.3.0 [120/1] via 12.1.13.3, 00:00:23, FastEthernet0/1
R 192.168.1.0/24 [120/1] via 12.1.12.2, 00:00:00, FastEthernet0/0

5. 相关知识

（1）RIP 路由的含义

这里，以其中的一条 RIP 路由"R 12.3.3.0 [120/1] via 12.1.13.3, 00:00:02, FastEthernet0/1"为例来理解 RIP 路由的含义。

- R：表示本条路由是经 RIP 协议产生的。
- 12.3.3.0：本条路由到达的目标网络。
- [120/1]：120 是 RIP 协议的默认管理距离；1 是度量值，在 RIP 中，从本路由器到目标网络中经过的路由器的个数，就是度量值，这里从 R1 到达 12.3.3.0 网络，经过了 R3 一个路由器，所以度量值为 1。
- Via：经过。
- 12.1.13.3：通告路由器的下一跳 IP 地址，指 12.3.3.0 网络是经过此 IP 的端口传来的。
- 00:00:02：从上次更新到查看路由表的时间。
- FastEthernet0/1：接收该路由的本路由器端口。

（2）子网路由问题

RIPv1 是主类路由协议，它能在主类网络边界自动汇总成**主类网络**，这样可以减少路由更新的带宽占用。但从上面的路由表可以看出，有四条 RIP 路由，并且前面三条是**子网路由**。当 RIPv1 满足整个网络所有地址属于同一个主类网络，且子网掩码长度相同时，就可以传递子网路由了。

在本实验中整个网络都属 12.0.0.0 的子网，且都是 24 位子网。

（3）等价路由

在 R1 的路由表输出中有：

R 12.1.23.0 [120/1] via 12.1.13.3, 00:00:02, FastEthernet0/1
     [120/1] via 12.1.12.2, 00:00:08, FastEthernet0/0

该路由到达的目标网络为 12.1.23.0，但有两条路径，度量值都为 1，这样的路由称为等价路由。

有了等价路由，就可以实现负载均衡。负载均衡有两种方式：一种是基于目的网络的负载均衡，称为快速交换；另一种是基于数据包的负载均衡，称为进程交换。在默认情况下采用的是快速交换：数据包在从源网络传递到目的网络时，只有一条到达路由被缓存，所有的数据包都使用相同的路径。

可使用命令 no ip cef 在全局模式下关闭快速交换：R1（config）#no ip cef，这样就使用了进程交换，此方式下路由器将基于每个数据包交替使用路径到达目的网络。

在默认情况下，RIP 最多支持四条等价路由实现负载均衡，可以通过下面的命令修改支持等价路由的条数：

  R1(config-router)#maximun-path *number-paths*

（4）被动接口

在路由器 R2 的 e0/0/0 上，连接的网络上没有 RIP 设备，但 R1 并不知道，仍每 30s 就向该接口发送一次更新，在这种情况下应将该接口设为被动接口，使之不再向该接口发送路由更新。

在没有设置为被动接口时，启用"debug ip rip"来查看 RIP 路由更新：

  R2#debug ip rip
  RIP protocol debugging is on
  ……

  RIP: sending v1 update to **255.255.255.255** via Ethernet0/0/0 (192.168.1.1)

  ……

可见，R2 正以**广播方式**向 Ethernet0/0/0 (192.168.1.1) 发送 RIP 更新。

将 e0/0/0 端口设为被动接口：

  R2(config)#router rip
  R2(config-router)#passive-interface e0/0/0

然后，再使用"R2#debug ip rip"查看 RIP 路由更新，此时不会再有向 e0/0/0 端口发送的路由更新了。

注意，最后需要用"R2#no debug all"关闭 debug 调试，在真实设备上，开启 debug 命令后会消耗大量的硬件资源。

实验 9 RIPv2

一、实验要求

- 掌握 RIPv2 的配置方法；
- 掌握 RIPv1 和 RIPv2 的区别；
- 理解 RIP 路由表含义；
- 掌握 auto-summary 的作用。

二、实验说明

RIPv2 同 RIPv1 一样，都是距离失量路由协议，用跳数作为度量值，最大跳数是 15 跳，同时具有相同的防环机制，采用每 30s 一次的周期更新，每次也是发送整张路由表。

但是，RIPv2 是 RIPv1 的增强版，具有一些增强特性：在路由更新中携带有子网掩码，可支持 VLSM、CIDR 和不连续子网；提供身份验证功能，包括 text 验证和 MD5 验证；采用组播更新取代了 RIPv1 中的广播更新，使路由更新效率更高；在自动汇总方面，RIPv1 默认的是自动汇总，并且不可手动关闭自动汇总，而 RIPv2 默认的是自动汇总，但是可以手动关闭自动汇总，并支持手动汇总。

关于 RIPv2 的这些特性，将在本实验和后面的实验中逐一演示。

三、实验拓扑

本实验使用的拓扑图如图 9-1 所示。

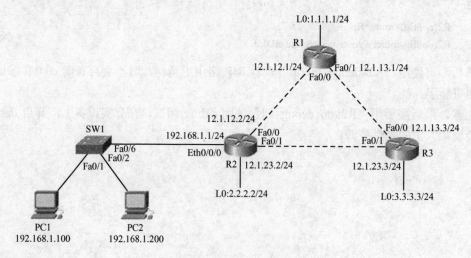

图 9-1 RIPv2 实验拓扑

四、实验配置

1. 路由器 1 的基本配置过程

```
Router>en
Router#conf t
Router(config)#host R1
R1(config)#int f0/0
R1(config-if)#ip add 12.1.12.1 255.255.255.0
R1(config-if)#no shut
R1(config-if)#int f0/1
R1(config-if)#ip add 12.1.13.1 255.255.255.0
R1(config-if)#no shut
R1(config)#int L0
R1(config-if)#ip add 1.1.1.1 255.255.255.0
R1(config-if)#exit
R1(config)#router rip
R1(config-if)#router rip              //配置 RIP 协议
R1(config-router)#ver 2               //指定 RIP 版本为 2
R1(config-router)#netw 12.0.0.0       //宣告路由时，尽管 RIPv2 支持子网，但宣告仍是主类网络地址
R1(config-router)#netw 1.0.0.0
R1(config-router)#no auto-summary     //关闭自动汇总，在主类网络边界不自动汇总
R1(config-router)#end
```

2. 路由器 2 的基本配置过程

```
Router>en
Router#conf t
Router(config)#host R1
R2(config)#int f0/0
R2(config-if)#ip add 12.1.12.2 255.255.255.0
R2 (config-if)#no shut
R2 (config-if)#int f0/1
R2 (config-if)#ip add 12.1.23.2 255.255.255.0
R2 (config-if)#no shut
R2(config-if)#int e0/0/0
R2(config-if)#ip add 192.168.1.1 255.255.255.0
R2(config-if)#no shut
R2(config)#int L0
R2(config-if)#ip add 12.2.2.2 255.255.255.0
R2(config-if)#exit
R2(config)#router rip
R2(config-router)#ver 2
R2(config-router)#netw 12.0.0.0
R2(config-router)#netw 2.0.0.0
R2(config-router)#netw 192.168.1.0
```

R2(config-router)#no auto-summary
R2(config-router)#end

3. 路由器 3 的基本配置过程

```
Router>en
Router#conf t
Router(config)#host R3
R3(config)#int f0/0
R3(config-if)#ip add 12.1.13.3 255.255.255.0
R3(config-if)#no shut
R3(config-if)#int f0/1
R3(config-if)#ip add 12.1.23.3 255.255.255.0
R3(config-if)#no shut
R3(config)#int L0
R3(config-if)#ip add 12.3.3.3 255.255.255.0
R3(config-if)#exit
R3(config)#router rip
R3(config-router)#ver 2
R3(config-router)#netw 3.0.0.0
R3(config-router)#netw 12.0.0.0
R3(config-router)#no auto-summary
R3(config-router)#end
```

4. 查看路由器 R1 的路由表

```
R1#show ip route
……
     1.0.0.0/24 is subnetted, 1 subnets
C       1.1.1.0 is directly connected, Loopback0
     2.0.0.0/24 is subnetted, 1 subnets
            //父路由，"is subnetted"表示有子网，/24 表示下面有 24 位长的子路
            //由，1 subnets 表示子网个数是 1；如果"is variably subnetted"，表示有
            //两个以上的掩码子网，且掩码长度不同
R       2.2.2.0 [120/1] via 12.1.12.2, 00:00:20, FastEthernet0/0
            //子路由，紧跟父路由之后
     3.0.0.0/24 is subnetted, 1 subnets
R       3.3.3.0 [120/1] via 12.1.13.3, 00:00:09, FastEthernet0/1
     12.0.0.0/24 is subnetted, 3 subnets              //3 subnets 表示有三个子网
C       12.1.12.0 is directly connected, FastEthernet0/0
C       12.1.13.0 is directly connected, FastEthernet0/1
R       12.1.23.0 [120/1] via 12.1.13.3, 00:00:09, FastEthernet0/1
                  [120/1] via 12.1.12.2, 00:00:20, FastEthernet0/0
R    192.168.1.0/24 [120/1] via 12.1.12.2, 00:00:20, FastEthernet0/0
```

可见，RIPv2 的所有路由条目更新都是带子网信息的。

5. 打开 auto-summary

现在在各路由器上，都打开自动汇总，然后查看路由表的变化（以 R1 为例）。

```
R1(config)#router rip
R1(config-router)#auto-summary
R1(config-router)#end
R1#show ip route
        1.0.0.0/24 is subnetted, 1 subnets
C       1.1.1.0 is directly connected, Loopback0
R       2.0.0.0/8 [120/1] via 12.1.12.2, 00:00:22, FastEthernet0/0
R       3.0.0.0/8 [120/1] via 12.1.13.3, 00:00:15, FastEthernet0/1
        //可见，打开自动汇总后，路由器在主类网络边界进行自动汇总，不产生子路由
        12.0.0.0/24 is subnetted, 3 subnets
C       12.1.12.0 is directly connected, FastEthernet0/0
C       12.1.13.0 is directly connected, FastEthernet0/1
R       12.1.23.0 [120/1] via 12.1.12.2, 00:00:22, FastEthernet0/0
                  [120/1] via 12.1.13.3, 00:00:15, FastEthernet0/1
R       192.168.1.0/24 [120/1] via 12.1.12.2, 00:00:22, FastEthernet0/0
```

6. 查看 RIPv2 路由更新过程

```
R1(config)#router rip
R1(config-router)#no auto-summary        //先关闭自动汇总
R1#debug ip rip
RIP protocol debugging is on
RIP: received v2 update from 12.1.13.3 on FastEthernet0/1
        //以下四行表示从 f0/1 端口收到 RIPv2 路由更新，含子网掩码长度和跳数，其中的
        //0.0.0.0 表示本路由器收到的更新
     2.2.2.0/24 via 0.0.0.0 in 2 hops
     3.3.3.0/24 via 0.0.0.0 in 1 hops
     12.1.23.0/24 via 0.0.0.0 in 1 hops
     192.168.1.0/24 via 0.0.0.0 in 2 hops
RIP: received v2 update from 12.1.12.2 on FastEthernet0/0
     2.2.2.0/24 via 0.0.0.0 in 1 hops
     3.3.3.0/24 via 0.0.0.0 in 2 hops
     12.1.23.0/24 via 0.0.0.0 in 1 hops
     192.168.1.0/24 via 0.0.0.0 in 1 hops
RIP: sending   v2 update to 224.0.0.9 via FastEthernet0/0 (12.1.12.1)
        //以下三行表示以组播方式（组播地址：224.0.0.9）从 f0/0 发出 RIPv2 更新，其中的
        //0.0.0.0 表示从本路由器发出的更新
RIP: build update entries
     1.1.1.0/24 via 0.0.0.0, metric 1, tag 0
     3.3.3.0/24 via 0.0.0.0, metric 2, tag 0
     12.1.13.0/24 via 0.0.0.0, metric 1, tag 0
……
```

7. 查看路由协议

```
R1#show ip protocols
Routing Protocol is "rip"
Sending updates every 30 seconds, next due in 3 seconds
```

```
Invalid after 180 seconds, hold down 180, flushed after 240
Outgoing update filter list for all interfaces is not set
Incoming update filter list for all interfaces is not set
Redistributing: rip
Default version control: send version 2, receive 2
                    //RIPv2 中，发和收都只是版本 2 的路由更新，从下面也可见到
  Interface              Send   Recv   Triggered RIP   Key-chain
  FastEthernet0/0         2      2
  FastEthernet0/1         2      2
  Loopback0               2      2
Automatic network summarization is in effect
Maximum path: 4
Routing for Networks:
    1.0.0.0
    12.0.0.0
Passive Interface(s):
Routing Information Sources:
    Gateway           Distance        Last Update
    12.1.12.2           120           00:00:05
    12.1.13.3           120           00:00:11
Distance: (default is 120)
```

其余与 RIPv1 是一样的。

可以通过命令来控制路由器的某个端口接收或发送 RIP 的版本，例如：

```
R1(config)#int f0/0
R1(config-if)#ip rip send version 1 2       //在 f0/0 端口，可以发送版本 1 和 2 的路由更新
R1(config-if)#ip rip receive version 2      //只接收版本 2 的路由更新
```

实验 10　RIPv2 路由验证

一、实验要求

- 掌握 RIP 的 text 验证的配置方法；
- 掌握 RIP 的 MD5 验证的配置方法。

二、实验说明

RIPv1 没有验证的功能，RIPv2 可以支持验证，并且有明文的 text 和 MD5 两种验证，通过验证功能，能保障路由器之间的可靠更新。

text 验证将会在网上以明文方式发送验证密码，而 MD5 模式则采用密文方式，安全性比 text 模式更高（本实验需要使用 DynamipsGUI 来完成）。

其配置过程包括：

① 定义钥匙链。
② 在钥匙链上定义一个或者一组钥匙。
③ 在接口上启用认证并指定使用的钥匙链。

三、实验拓扑

实验拓扑如图 10-1 所示。

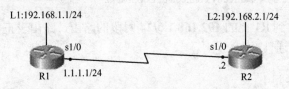

图 10-1　RIPv2 路由验证实验拓扑

四、实验过程

在 R1 上配置 text 验证模式（后面再讲 MD5 验证模式）：

```
Router#conf t                    //前面这部分是基本配置
Router(config)#host R1
```

```
R1(config)#conf t
R1(config)#int s1/0
R1(config-if)#ip add 1.1.1.1 255.255.255.0
R1(config-if)#no shut
R1(config-if)#int L1
R1(config-if)#ip add 192.168.1.1 255.255.255.0
R1(config-if)#router rip
R1(config-router)#ver 2
R1(config-router)#netw 1.0.0.0
R1(config-router)#netw 192.168.1.0
R1(config-router)#exit
R1(config)#key chain abc                          //创建密钥链，取名 abc，在 R2 上可以用不同密钥链名
R1(config-keychain)#key 1                         //配第一把钥匙，key 1
R1(config-keychain-key)#key-string cisco          //配置密码：cisco，两端要求一致
R1(config-keychain-key)#int s1/0                  //在 s1/0 端口上启用认证
R1(config-if)#ip rip authentication mode text     //验证模式是 text 模式，此模式的密码在网上以明文
                                                  //形式发送，安全性不高
R1(config-if)#ip rip authentication key-chain abc //在 s1/0 端口上调用的验证密钥链是前面所创建的密
                                                  //钥链 abc
```

在 R2 上，按上述相同配置方法完成配置。

五、实验总结

查看 R2 的路由表：

```
R2#show ip route
......
     1.0.0.0/24 is subnetted, 1 subnets
C    1.1.1.0 is directly connected, Serial1/0
R    192.168.1.0/24 [120/1] via 1.1.1.1, 00:01:40, Serial1/0
C    192.168.2.0/24 is directly connected, Loopback2
```

可见，R2 上已学到 R1 上的 **192.168.1.0/24** 网段的路由，路由表正常。
再用 ping 测试连通性：

```
R2#ping 192.168.1.1

Type escape sequence to abort.
Sending 5, 100-byte ICMP Echos to 192.168.1.1, timeout is 2 seconds:
!!!!!
Success rate is 100 percent (5/5), round-trip min/avg/max = 4/35/68 ms
```

可见，配置正常。（在此，读者可以修改一下两端配置的密码，如一端是 cisco，另一端是 ccna，再来查看路由表以及 ping 测试，看结果如何，为什么？）
下面证实 text 模式在网上是以明文方式发送密码的。在 R2 上开启 debug 功能：

```
R2#debug ip rip
RIP protocol debugging is on
R2# address 1.1.1.2 255.255.255.0
*Dec 13 14:46:46.691:RIP:received packet with text authentication cisco
```
认证密码：cisco

从上可见，从对端接口 1.1.1.2 上收到的验证密码就是所配置的密码：cisco。

现在把 text 验证模式改为 MD5 模式：

```
R1(config-if)#ip rip authentication mode md5
```

同样，在 R2 上也改为 MD5 后，开启 dubug 功能，查看从对端接口中收到的验证密码情况：

```
R2#debug ip rip
……
*Dec 13 15:47:04.347: RIP: received packet with MD5 authentication
……
```

可见，从 debug 的输出中，已看不到口令了，从而防止了口令被非法获取，提高了安全性。

实验 11　向 RIP 注入默认路由

一、实验要求

- 了解注入默认路由的作用；
- 掌握向 RIP 网络中注入默认路由的方法。

二、实验说明

配置在企业网边界路由器上的默认路由可简化企业网内部路由器的配置。由图 11-1 可见，R1 是企业边界路由器，内网配置了 RIP 路由协议，通过在 R1 上宣告一条动态的默认路由，在内部路由器 R2 和 R3 上就不需要配置去往边界路由器的默认路由了，从而减少了内网配置的复杂性。

本实验可在 Cisco Packet Tracer 模拟软件中实现完整配置。

三、实验拓扑

实验拓扑如图 11-1 所示。

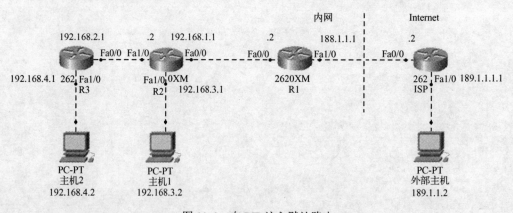

图 11-1　向 RIP 注入默认路由

四、实验过程

在本实验中，只需对 ISP 完成基本配置即可。

1. 配置 ISP 路由器

```
Router#conf t
R3(config)#host ISP
ISP(config)#int f0/0
ISP(config-if)#ip add 188.1.1.2 255.255.255.0
ISP(config-if)#int f1/0
ISP(config-if)#ip add 189.1.1.1 255.255.255.0
ISP(config-if)#no shut
```

2. 配置边界路由器 R1

```
Router#conf t
Router(config)#host R1
R1(config)#int f0/0
R1(config-if)#ip add 192.168.1.2 255.255.255.0      //先按照拓扑图，配置各接口 IP 地址
R1(config-if)#no shut
R1(config-if)#int f1/0
R1(config-if)#ip add 188.1.1.1 255.255.255.0
R1(config-if)#no shut
R1(config-if)#exit
R1(config)#router rip
R1(config-router)#netw 192.168.1.0                   //只公告内部网络，不能公告与外网相连的网络
R1(config-router)#default-information originate      //声明 R1 是默认路由起源，R1 则会向其他 RIP
                                                     //路由器宣告自己是默认路由
R1(config-router)#exit
R1(config)#ip route 0.0.0.0 0.0.0.0 188.1.1.2        //配置默认路由，实现对 Internet 的访问
R1(config)#int f0/0
R1(config-if)#ip nat inside                          //指定 PAT 地址转换的内外端口（关于网络地址
                                                     //转换，后面章节会专门讲述）
R1(config-if)#int f1/0
R1(config-if)#ip nat outside
R1(config-if)#exit
R1(config)#access-list 1 permit 192.168.1.0 0.0.0.255   //定义访问控制列表（关于 ACL 在后面章节讲述）
R1(config)#access-list 1 permit 192.168.2.0 0.0.0.255   //将列表中的所有网络通过端口 f1/0 转发到外网
R1(config)#access-list 1 permit 192.168.3.0 0.0.0.255   //这四条可以缩写成一条：
                                                        //access-list 1 permit 192.168.0.0 0.0.255.255
R1(config)#access-list 1 permit 192.168.4.0 0.0.0.255
R1(config)#ip nat inside source list 1 int f1/0 overload
R1(config)#
```

3. 配置内网路由器 R2

```
Router(config)#host R2
R2(config)#int f0/0
R2(config-if)#ip add 192.168.1.1 255.255.255.0
R2(config-if)#no shut
R2(config-if)#int f1/1
```

```
R2(config-if)#ip add 192.168.2.2 255.255.255.0
R2(config-if)#no shut
R2(config-if)#int f1/0
R2(config-if)#ip add 192.168.3.1 255.255.255.0
R2(config-if)#no shut
R2(config-if)#exit
R2(config)#router rip
R2(config-router)#netw 192.168.1.0
R2(config-router)#netw 192.168.2.0
R2(config-router)#netw 192.168.3.0
```

4. 配置内网路由器 R3

```
Router#conf t
Router(config)#host R3
R3(config)#int f1/0
R3(config-if)#ip add 192.168.4.1 255.255.255.0
R3(config-if)#no shut
R3(config-if)#int f0/0
R3(config-if)#ip add 192.168.2.1 255.255.255.0
R3(config-if)#no shut
R3(config-if)#exit
R3(config)#router rip
R3(config-router)#netw 192.168.2.0
R3(config-router)#netw 192.168.4.0
```

内网路由器 R2 和 R3 的配置与基本的 RIP 路由协议完全相同。

五、实验总结

用内网主机 2（或主机 1）去测试到达外网主机的连通性，如图 11-2 所示。

```
PC>ping 189.1.1.2

Pinging 189.1.1.2 with 32 bytes of data:

Reply from 189.1.1.2: bytes=32 time=39ms TTL=125
Reply from 189.1.1.2: bytes=32 time=7ms TTL=125
Reply from 189.1.1.2: bytes=32 time=13ms TTL=125
Reply from 189.1.1.2: bytes=32 time=20ms TTL=125

Ping statistics for 189.1.1.2:
    Packets: Sent = 4, Received = 4, Lost = 0 (0% loss),
Approximate round trip times in milli-seconds:
    Minimum = 7ms, Maximum = 39ms, Average = 19ms
```

图 11-2　测试到达外网主机的连通性

可见，从内网主机能访问外网主机。

查看边界路由器 R1 的路由表：

```
R1#show ip route
……                    //此处省掉的是 Codes，即代码说明部分，以省篇幅，后同
     188.1.0.0/24 is subnetted, 1 subnets
C       188.1.1.0 is directly connected, FastEthernet1/0
C       192.168.1.0/24 is directly connected, FastEthernet0/0
R       192.168.2.0/24 [120/1] via 192.168.1.1, 00:00:02, FastEthernet0/0
R       192.168.3.0/24 [120/1] via 192.168.1.1, 00:00:02, FastEthernet0/0
R       192.168.4.0/24 [120/2] via 192.168.1.1, 00:00:02, FastEthernet0/0
S*      0.0.0.0/0 [1/0] via 188.1.1.2
```

可见，R1 上有一条默认路由到 ISP 的入口，它的作用是将内网中转发来的所有未知网络，转到 ISP 的接入路由器。这里需注意，不能将与 ISP 相连的接口也公告出来，否则可能导致内网被外部人员非法访问。

再查看内网路由器 R3 的路由表：

```
R3#show ip route
……
R       192.168.1.0/24 [120/1] via 192.168.2.2, 00:00:16, FastEthernet0/0
C       192.168.2.0/24 is directly connected, FastEthernet0/0
R       192.168.3.0/24 [120/1] via 192.168.2.2, 00:00:16, FastEthernet0/0
C       192.168.4.0/24 is directly connected, FastEthernet1/0
R*      0.0.0.0/0 [120/2] via 192.168.2.2, 00:00:16, FastEthernet0/0
```

可见，R3 学到一条"R*   0.0.0.0/0 [120/2] via 192.168.2.2, 00:00:16, FastEthernet0/0"路由，这是一条通过 RIP 路由协议学到的动态默认路由（并没有在 R3 配置这条路由）。

实验 12　EIGRP 的配置

一、实验要求

- 掌握 EIGRP 的基本配置方法；
- 理解 EIGRP 中 FD、AD 的计算方法；
- 理解 EIGRP 三张表的相关内容；
- 掌握 EIGRP 负载均衡的相关内容。

二、实验说明

EIGRP 以前属于 Cisco 私有路由协议，但在 2013 年已公有化了，因此 EIGRP 在网络中的应用有所增加。本实验是 EIGRP 的基本配置，主要是掌握 EIGRP 的一些常规属性，为后面进一步深入学习 EIGRP 路由协议和工程应用打下基础。

三、实验拓扑

本实验使用的拓扑图如图 12-1 所示。

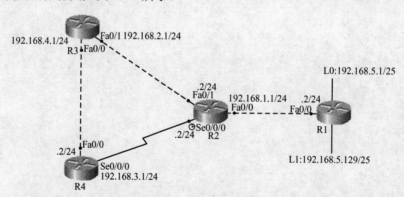

图 12-1　本实验拓扑图

四、实验配置

1. R1 的配置过程

R1(config)#int L0

```
R1(config-if)#ip add 192.168.5.1 255.255.255.128      //基本配置
R1(config-if)#int L1
R1(config-if)#ip add 192.168.5.129 255.255.255.128
R1(config-if)#int f0/0
R1(config-if)#ip add 192.168.1.2 255.255.255.0
R1(config-if)#no shut
R1(config-if)#exit
R1(config)#router eigrp 100           //启用 EIGRP 进程，AS 号为 100，在同一个互联网络中所有
                                      //路由器 AS 号相同才能相互学习路由
R1(config-router)#netw 192.168.5.0    //这里 192.168.5.0 后没加上反掩码，表示凡是属于该有类网
                                      //络地址的所有接口都将启用 EIGRP，这里包括 L0 和 L1 两
                                      //个逻辑接口（如果仅打算公告一个子网，则需加上反掩码）
R1(config-router)#netw 192.168.1.0 0.0.0.255
R1(config-router)#auto-summary        //配置自动汇总，可以不加此语句，因为 EIGRP 默认在主类
                                      //网络边界自动汇总
R1(config-router)#end
```

2. R2 的配置过程

```
R2(config)#router eigrp 100
R2(config-router)#auto-summary
R2(config-router)#netw 192.168.1.0 0.0.0.255    //常规格式的配置都需要加上反掩码，以指明具体的
                                                //网络或子网
R2(config-router)#netw 192.168.2.0 0.0.0.255
R2(config-router)#netw 192.168.3.0 0.0.0.255
R2(config-router)#end
```

3. R3 的配置过程

```
R3(config)#router eigrp 100
R3(config-router)#auto-summary
R3(config-router)#netw 192.168.2.0 0.0.0.255
R3(config-router)#netw 192.168.4.0 0.0.0.255
R3(config-router)#end
```

4. R4 的配置过程

```
R4(config)#router eigrp 100
R4(config-router)#auto-summary
R4(config-router)#netw 192.168.4.0 0.0.0.255
R4(config-router)#netw 192.168.3.0 0.0.0.255
R4(config-router)#end
```

五、实验总结

1. EIGRP 中的相关概念

Successor（后继）：进入路由表的路由中，到达目标网络的下一个路由器。

FD（Feasible Distance，可行距离）：到达目标网络的最小度量值。

AD（Advertised Distance，通告距离）：又称为 RD（Reported Distance，报告距离），指下一跳路由器通告到目标网络的度量值。这里的下一跳路由器，不一定是后继路由器。

FS（Feasible Successor，可行后继）：后继的备份路由器，FS 也是本路由器的邻居路由器。DUAL 算法能快速收敛的原因就是在拓扑表中保存有到达目标网络的替代路径，当网络拓扑发生变化时，DUAL 不需要重新计算，而直接将 Feasible Successor 变为 Successor。

FC（Feasibility Condition，可行条件）：可行条件 FC 是指一个路由器要成为可行后继的条件是 AD<FD（AD 是下一跳路由器到达目标网络的度量值，FD 是本路由器到达目标网络的最小度量值）。

2．关于 FD、AD 值的计算

EIGRP 度量值 Metric 的计算公式：

Metric=256*{K1（10^7/带宽）+K2（10^7/带宽）/（256-负载）+K3*延迟+K5/(可靠性+K4)}

默认情况下，K1=K3=1，其他的 K 值都是 0，因此，EIGRP 度量值简化为：

Metric =256*（10^7/最小带宽+累积延时/10）

在此以计算 R2 到达网络 192.168.4.0 为例，来说明度量值计算方法。

首先需要查看接口带宽和延时信息，这里先查看 R2 上 f0/1 端口：

> R2#show int f0/1
> FastEthernet0/1 is up, line protocol is up (connected)
>   Hardware is Lance, address is 00e0.f9aa.8602 (bia 00e0.f9aa.8602)
>   Internet address is 192.168.2.2/24
>   MTU 1500 bytes, **BW 100000 kbit, DLY 100 usec,**
>   ……

这里的 **BW 100000 kbit, DLY 100 usec**，就是端口带宽和端口延时，在计算 Metric 时，带宽采用的是从本路由器到目标网络的**最小带宽**，端口延时采用的是以本路由器到目标网络方向的外出端口的延时。

同样，也可以查看 R2 上 s0/0/0 端口：

> R2#show int s0/0/0
> Serial0/0/0 is up, line protocol is up (connected)
>   Hardware is HD64570
>   Internet address is 192.168.3.2/24
>   MTU 1500 bytes, **BW 1544 kbit, DLY 20000 usec,**
>   ……

从 R2 到目标网络 192.168.4.0 有两条路径：一条是以太网端口，一条是串行端口。其度量值分别计算如下。

- 从以太网端口到目标网络：Metric=256*(10000000/100000+(100+100)/10)=30720
- 从串行端口到目标网络：Metric=256*(10000000/1544+(100+20000)/10)= 2172416

FD 就是从本路由器 R2 开始，到达目标网络的度量值，上面计算出的这两个 Metric 值就是从 R2 到目标网络 192.168.4.0 的两条不同路径的 FD 值。

AD 值是从本路由器 R2 的下一个路由器开始，到达目标网络的度量值，在本例中，R2

到目标网络有两条路径，一条经过路由器 R3，另一条经过路由器 R4，求 AD 值就是求 R3 和 R4 到目标网络的 Metric 值，通过此拓扑图可见，这两个 AD 值是一样的：

Metric=256*(10000000/100000+100/10)=28160，即 AD 值为 28160。

这里计算的 FD 值和 AD 值，在后面查看拓扑表时能够得到证实。

3．查看邻居表

在路由器 R2 上查看邻居表：

```
R2#show ip eigrp neighbors
IP-EIGRP neighbors for process 100
   H   Address         Interface    Hold Uptime     SRTT   RTO    Q    Seq
                                    (sec)           (ms)          Cnt  Num
   0   192.168.2.1     Fa0/1        13   00:22:28   40     1000   0    30145
   1   192.168.3.1     Se0/0/0      10   00:22:28   40     1000   0    30159
   2   192.168.1.2     Fa0/0        13   00:17:11   40     1000   0    15070
```

从邻居表中可以看到，R2 与哪些端口（显示的是对端的 IP 地址）建立了邻居关系，是通过本路由器的哪些端口来建立的。例如，R2 通过本路由器端口 f0/1 与 IP 地址为 192.168.2.1 的端口建立了邻居关系。

Hold：保持时间，还有 13 秒钟（正常情况下这个时间是 14 递减到 10，在收到一个 hello 包后，又从 14 开始递减，如果最后变成 0 时邻居关系就被取消）。

Uptime：邻居路由器进入邻居表的时间；

其他一些相关信息在此不再一一讨论。

4．查看路由表

在路由器 R2 上查看路由表：

```
R2#show ip route
Codes: C - connected, S - static, I - IGRP, R - RIP, M - mobile, B - BGP
       D - EIGRP, EX - EIGRP external, O - OSPF, IA - OSPF inter area
       N1 - OSPF NSSA external type 1, N2 - OSPF NSSA external type 2
       E1 - OSPF external type 1, E2 - OSPF external type 2, E - EGP
       i - IS-IS, L1 - IS-IS level-1, L2 - IS-IS level-2, ia - IS-IS inter area
       * - candidate default, U - per-user static route, o - ODR
       P - periodic downloaded static route

Gateway of last resort is not set

C    192.168.1.0/24 is directly connected, FastEthernet0/0
C    192.168.2.0/24 is directly connected, FastEthernet0/1
C    192.168.3.0/24 is directly connected, Serial0/0/0
D    192.168.4.0/24 [90/30720] via 192.168.2.1, 00:00:50, FastEthernet0/1
D    192.168.5.0/24 [90/156160] via 192.168.1.2, 00:00:04, FastEthernet0/0
```

可见，通过 EIGRP 路由协议学习到的路由条目前面有字母"D"作为标识。以"**D 192.168.4.0/24 [90/30720] via 192.168.2.1, 00:00:50, FastEthernet0/1**"为例，192.168.4.0/24 是通过 EIGRP 学习到的目标网络；[90/30720]表示管理距离为 90，度量值为 30720；via

192.168.2.1 表示到达目标网络的下一跳路由器入口的 IP 地址；FastEthernet0/1 表示本路由器外出端口名称。

由于在 R1 上使用了"auto-summary"进行了自动汇总，因此并没有看到到达 192.168.5.0/25 和 192.168.5.128/25 的子网路由。现在在 R1 上关闭自动汇总：

```
R1(config)#route eigrp 100
R1(config-router)#no auto-summary
```

然后再到 R2 上查看路由表：

```
R2#show ip route
......           //此处省略代码部分

C    192.168.1.0/24 is directly connected, FastEthernet0/0
C    192.168.2.0/24 is directly connected, FastEthernet0/1
C    192.168.3.0/24 is directly connected, Serial0/0/0
D    192.168.4.0/24 [90/30720] via 192.168.2.1, 00:05:22, FastEthernet0/1
     192.168.5.0/24 is variably subnetted, 3 subnets, 2 masks
D    192.168.5.0/24 is a summary, 00:00:06, Null0
D    192.168.5.0/25 [90/156160] via 192.168.1.2, 00:00:06, FastEthernet0/0
D    192.168.5.128/25 [90/156160] via 192.168.1.2, 00:00:06, FastEthernet0/0
```

其中：

"**192.168.5.0/24 is variably subnetted, 3 subnets, 2 masks**"是一条父路由，不是最终路由，声明使用了变长子网掩码，有3个子网和2个子网掩码。

"**D   192.168.5.0/24 is a summary, 00:00:06, Null0**"是一条汇总路由，外出接口指向"Null0"，即空接口，发往空接口的数据包将丢弃，用于避免路由环路。

"**D   192.168.5.0/25 [90/156160] via 192.168.1.2, 00:00:06, FastEthernet0/0**
D   192.168.5.128/25 [90/156160] via 192.168.1.2, 00:00:06, FastEthernet0/0"
这两条是由于关闭了自动汇总，而从 R1 上学来的子路由。

5. 查看拓扑表

```
R2#show ip eigrp topology
IP-EIGRP Topology Table for AS 100

Codes: P - Passive, A - Active, U - Update, Q - Query, R - Reply,
       r - Reply status

P 192.168.1.0/24, 1 successors, FD is 28160
        via Connected, FastEthernet0/0
P 192.168.2.0/24, 1 successors, FD is 28160
        via Connected, FastEthernet0/1
P 192.168.3.0/24, 1 successors, FD is 2169856
        via Connected, Serial0/0/0
P 192.168.4.0/24, 1 successors, FD is 30720
        via 192.168.2.1 (30720/28160), FastEthernet0/1
```

                via 192.168.3.1 (2172416/28160), Serial0/0/0
P 192.168.5.0/24, 1 successors, FD is 156160
                via Summary (156160/0), Null0
P 192.168.5.0/25, 1 successors, FD is 156160
                via 192.168.1.2 (156160/128256), FastEthernet0/0
P 192.168.5.128/25, 1 successors, FD is 156160
                via 192.168.1.2 (156160/128256), FastEthernet0/0

下面以"**P 192.168.4.0/24, 1 successors, FD is 30720**
    via 192.168.2.1 (30720/28160), FastEthernet0/1
    via 192.168.3.1 (2172416/28160), Serial0/0/0"为例，来介绍拓扑表。

从"Codes"部分可知，P- Passive 表示被动路由。R2 的拓扑表中，所有路由都是被动路由，表示路由处于稳定状态，而其余几种状态由于 EIGRP 收敛速度非常快，一般在查看拓扑表时不会看到它们的存在。

"**192.168.4.0/24, 1 successors, FD is 30720**"表示从 R2 到达的目标网络，有一个后继（**1 successors**），其 FD 值是 **30720**，此值与"**via 192.168.2.1 (30720/28160), FastEthernet0/1**"路由中的 FD 相同，因此 192.168.2.1 就是 R2 到达目标网络 192.168.4.0 的后继路由，这条路由既可进入拓扑表，又可进入路由表（可以对比上面 R2 的路由表来证实）。

"**via 192.168.3.1 (2172416/28160), Serial0/0/0**"中的 FD 值是 **2172416**，比前一条路由的 FD 值大，不能进入路由表，但由于其 AD<FD，即 **286160<30720**，满足可行条件，因此可以进入拓扑表中，192.168.3.1 也就成了 R2 到达目标网络 192.168.4.0 的可行后继路由。

6. 负载均衡

在查看 R2 的拓扑表时，可以看到有两条路由能到达 192.168.4.0/24 网络（也可以通过观察图 12-1 得知从 R2 到 192.168.4.0/24 网络有两条路径），这两条路由的 FD 值是不等的，分别是 2172416 和 30720。

EIGRP 既支持等价负载均衡，又支持不等价负载均衡。如果 FD 值不等，就只能使用不等价负载均衡。

在查看 R2 的路由表时，只有一条到达 192.168.4.0/24 网络的路由条目，在此可通过配置比例因子，使 R2 拓扑表中能到达 192.168.4.0/24 网络的两条路径都成为路由表的条目。

比例因子的计算方法如下：用拓扑表中的两条路径的 FD 值相除，即 2172416/30720，并向上取整，得到比例因子为 71。

        R2(config)#router eigrp 100
        R2(config-router)#variance 71

在默认情况下，EIGRP 采用的是等价负载均衡，其 variance 值为 1，variance 的取值范围是 1～128，配上该范围内比例因子后，满足可行条件链路都将被放入路由表中，下面再次查看 R2 的路由表：

        R2#show ip route
        ……

        C   192.168.1.0/24 is directly connected, FastEthernet0/0

	C	192.168.2.0/24 is directly connected, FastEthernet0/1
	C	192.168.3.0/24 is directly connected, Serial0/0/0
	D	**192.168.4.0/24 [90/30720] via 192.168.2.1, 00:00:06, FastEthernet0/1**
		[90/2172416] via 192.168.3.1, 00:00:07, Serial0/0/0
		192.168.5.0/24 is variably subnetted, 3 subnets, 2 masks
	D	192.168.5.0/24 is a summary, 00:00:06, Null0
	D	192.168.5.0/25 [90/156160] via 192.168.1.2, 00:00:06, FastEthernet0/0
	D	192.168.5.128/25 [90/156160] via 192.168.1.2, 00:00:06, FastEthernet0/0

可见，已有两条非等价路由可到达 192.168.4.0/24 网络，这两条路由是按配置的比例因子进行负载均衡的。

实验 13 向 EIGRP 注入默认路由

一、实验要求

- 了解注入默认路由的作用；
- 掌握向 EIGRP 网络中注入默认路由的方法。

二、实验说明

配置在企业网边界路由器上的默认路由可简化企业网内部路由器的配置。由图 13-1 可见，R1 是企业边界路由器，内网配置的是 EIGRP 路由协议，通过在 R1 上向内网注入一条动态的默认路由，在内部路由器 R2 和 R3 上就不需要配置去往边界路由器的默认路由了。本实验需要在 DynamipsGUI 下完成（在 Cisco Packet Tracer 模拟软件中不能正常实现部分功能）。

三、实验拓扑

本实验拓扑如图 13-1 所示。

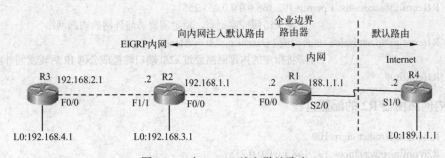

图 13-1 向 EIGRP 注入默认路由

四、实验配置

1. ISP 路由器的配置过程

```
Router#conf t
R3(config)#host ISP
ISP(config)#int S1/0
```

ISP(config-if)#ip add 188.1.1.2 255.255.255.0
ISP(config-if)#int L0
ISP(config-if)#ip add 189.1.1.1 255.255.255.0
ISP(config-if)#no shut

2. 边界路由器 R1 的配置过程

Router#conf t
Router(config)#host R1
R1(config)#int s2/0
R1(config-if)#ip add 188.1.1.1 255.255.255.0
R1(config-if)#no shut
R1(config-if)#int f0/0
R1(config-if)#ip add 192.168.1.2 255.255.255.0
R1(config-if)#no shut
R1(config-if)#exit
R1(config)#router eigrp 100
R1(config-router)#netw 192.168.1.0 0.0.0.255
R1(config-router)#redistribute static
R1(config-router)#exit
R1(config)#ip route 0.0.0.0 0.0.0.0 s2/0
　　　　　//配置一条默认路由指向外部网络，表示当边界路由器收到需要到达的目标网络在
　　　　　//其他路由条目中没有时，就使用此路由条目，将数据发给 ISP 路由器
R1(config)#int f0/0
R1(config-if)#ip nat inside　　//指定 f0/0 端口为内网端口
R1(config-if)#int s2/0
R1(config-if)#ip nat outside　　//指定 s2/0 端口为外网端口
R1(config-if)#exit
R1(config)#access-list 1 permit 192.168.0.0 0.0.255.255
　　　　　　　　　//定义访问控制列表，指定需要访问外网的内部网络
R1(config)#ip nat inside source list 1 interface s2/0 overload
　　　　　　　　//将指定的内部网络经过 s2/0 端口转换成公网 IP 后转发到外网
R1(config)#

3. 内网路由器 R2 的配置过程

R2(config)#router eigrp 100
R2(config-router)#netw 192.168.1.0 0.0.0.255
R2(config-router)#netw 192.168.2.0 0.0.0.255
R2(config-router)#netw 192.168.3.0 0.0.0.255
R2(config-router)#end

4. 内网路由器 R3 的配置过程

R3(config)#router eigrp 100
R3(config-router)#netw 192.168.2.0 0.0.0.255
R3(config-router)#netw 192.168.4.0 0.0.0.255
R3(config-router)#end

内网路由器 R2 和 R3 的配置与基本的 EIGRP 路由协议完全相同，它们收到需要访问外网的请求时，就使用边界路由器 R1 向它们注入的默认路由来进行转发。

五、实验总结

1. 连通性测试

从内部路由器 R3 上去 ping ISP 路由器的 L0 端口：

R3#ping 189.1.1.1 source 192.168.4.1

Type escape sequence to abort.
Sending 5, 100-byte ICMP Echos to 189.1.1.1, timeout is 2 seconds:
Packet sent with a source address of 192.168.4.1
!!!!!
Success rate is 100 percent (5/5), round-trip min/avg/max = 16/45/116 ms

可见，从内部网络能访问外部网络。

2. 查看边界路由器 R1 的路由表

R1#show ip route
......   //此处省掉的是 Codes，即代码说明部分，以节省篇幅，后同
      188.1.0.0/24 is subnetted, 1 subnets
C   188.1.1.0 is directly connected, Serial2/0
D   192.168.4.0/24 [90/158720] via 192.168.1.1, 01:03:56, FastEthernet0/0
C   192.168.1.0/24 is directly connected, FastEthernet0/0
D   192.168.2.0/24 [90/30720] via 192.168.1.1, 01:05:54, FastEthernet0/0
D   192.168.3.0/24 [90/156160] via 192.168.1.1, 01:05:48, FastEthernet0/0
S*   0.0.0.0/0 [1/0] via 188.1.1.2

可见，R1 上有一条默认路由到 ISP 的入口，它的作用是将内网中转发来的所有未知网络，转到 ISP 的接入路由器。这里需注意，不能将与 ISP 相连的接口也公告出来，这可能会导致内网被外部人员非法访问。

再查看内网路由器 R2 和 R3 的路由表：

R2#show ip route
......
D   192.168.4.0/24 [90/156160] via 192.168.2.1, 00:00:47, FastEthernet1/1
C   192.168.1.0/24 is directly connected, FastEthernet0/0
C   192.168.2.0/24 is directly connected, FastEthernet1/1
C   192.168.3.0/24 is directly connected, Loopback0
D*EX 0.0.0.0/0 [170/2172416] via 192.168.1.2, 00:02:49, FastEthernet0/0

R3#show ip route
......
C   192.168.4.0/24 is directly connected, Loopback0
D   192.168.1.0/24 [90/30720] via 192.168.2.2, 00:00:34, FastEthernet0/0
C   192.168.2.0/24 is directly connected, FastEthernet0/0

```
D       192.168.3.0/24 [90/156160] via 192.168.2.2, 00:00:34, FastEthernet0/0
D*EX    0.0.0.0/0 [170/2174976] via 192.168.2.2, 00:00:34, FastEthernet0/0
```

可见，在内网路由器 R2 和 R3 上，均自动产生了一条代码为"**D*EX**"的路由，这是通过重分布静态默认路由向 EIGRP 网络注入的默认路由，其中"**EX**"是通过重分布进入 EIGRP 网络的标识。

静态重分布是向 EIGRP 注入默认路由非常简单的一种方式。另外，还有两种方式可以向 EIGRP 注入默认路由。

3．使用通告默认路由的方式向 EIGRP 注入默认路由

```
R1(config)#router eigrp 100
R1(config-router)#no redistribute static     //删除静态重分布路由指令
R1(config-router)#netw 0.0.0.0               //通告默认路由的方式向 EIGRP 注入默认路由
```

查看 R2 和 R3 的路由表：

```
R2#show ip route
……
D       188.1.0.0/16 [90/2172416] via 192.168.1.2, 00:15:06, FastEthernet0/0
D       192.168.4.0/24 [90/156160] via 192.168.2.1, 01:34:42, FastEthernet1/1
C       192.168.1.0/24 is directly connected, FastEthernet0/0
C       192.168.2.0/24 is directly connected, FastEthernet1/1
C       192.168.3.0/24 is directly connected, Loopback0
D*      0.0.0.0/0 [90/2172416] via 192.168.1.2, 00:03:48, FastEthernet0/0
R3#show ip route
……
D       188.1.0.0/16 [90/2174976] via 192.168.2.2, 00:10:46, FastEthernet0/0
C       192.168.4.0/24 is directly connected, Loopback0
D       192.168.1.0/24 [90/30720] via 192.168.2.2, 00:10:46, FastEthernet0/0
C       192.168.2.0/24 is directly connected, FastEthernet0/0
D       192.168.3.0/24 [90/156160] via 192.168.2.2, 00:10:46, FastEthernet0/0
D*      0.0.0.0/0 [90/2174976] via 192.168.2.2, 00:00:06, FastEthernet0/0
```

可见，在内网路由器 R2 和 R3 上，均自动产生了一条默认路由，通过此默认路由（在路由器 R2、R3 上并未配置），可以实现对外部网络的访问。

4．使用"ip default-network"命令向 EIGRP 网络注入默认路由

这里简单介绍一下这种方式（接着上一种方式）：

```
R1(config)#router eigrp 100
R1(config-router)#no netw 0.0.0.0            //删除通告的默认路由
R1(config-router)#netw 188.1.0.0             //公告与外网相连的网络
R1(config-router)#exit
R1(config)#ip default-network 188.1.0.0
```

向 EIGRP 网络注入默认路由，由于"ip default-network"命令只支持有类 IP，所以，在使用这种方式的时候，应该使用该网段对应的主类网络号，这里是 B 类地址，其主类网络号就是 188.1.0.0。

实验 14 EIGRP 验证

一、实验要求

掌握 EIGRP 的 MD5 验证的配置方法。

二、实验说明

EIGRP 的验证方式只有 MD5 一种方式，不像 RIP 和 OSPF 那样，既支持明文验证，又支持 MD5 验证。EIGRP 的 MD5 验证配置步骤与 RIP 上的配置 MD5 验证类似（本实验需要使用 DynamipsGUI 来完成），其过程包括：

① 定义钥匙链；
② 在钥匙链上定义一个或者一组钥匙；
③ 在接口上启用认证并指定使用的钥匙链。

三、实验拓扑

配置 EIGRP 验证使用的拓扑如图 14-1 所示。

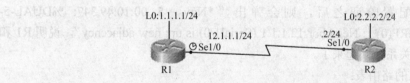

图 14-1 EIGRP 验证拓扑图

四、实验配置

首先根据图 14-1 所示，完成对 R1 和 R2 的基本配置以及 EIGRP 协议的配置，然后，在 R1 上配置验证：

```
R1(config)#key chain can                              //定义钥匙链，取名 cna
R1(config-keychain)#key 1                             //定义第一个密钥，可以定义多个
R1(config-keychain-key)#key-string abc                //密钥串（密码）为 abc
R1(config-keychain-key)#int s1/0                      //在串行接口 s1/0 上启用
R1(config-if)#ip authentication key-chain eigrp 1 can //调用前面定义的钥匙链 cna
R1(config-if)#ip authentication mode eigrp 1 md5      //使用 MD5 验证方式
```

再在 R2 上配置验证：

```
R2(config)#key chain xyz                              //此处名字可与 R1 不同
R2(config-keychain)#key 1
```

```
R2(config-keychain-key)#key-string abc              //要求与 R1 上的密钥串相同
R2(config-keychain)#int s1/0
R2(config-if)#ip authentication key-chain eigrp 1 xyz   //调用前面定义的钥匙链 xyz
R2(config-if)#ip authentication mode eigrp 1 md5
```

五、实验总结

在 EIGRP 验证的配置中，要求串行接口两端的密钥串（密码）相同才可完成验证，而钥匙链的名字可以不同。

如果仅在 R1 配置了验证，而 R2 上没有配置，则会弹出"*Nov 5 00:10:47.267: %DUAL-5-NBRCHANGE: IP-EIGRP(0) 1: Neighbor 12.1.1.2 (Serial1/0) is down: Interface Goodbye received"，这说明 R1 和 R2 间验证失败，需要在 R2 上也配置验证。

此时查看 R1 的路由表：

```
R1#show ip route
……
         1.0.0.0/8 is variably subnetted, 2 subnets, 2 masks
C        1.1.1.0/24 is directly connected, Loopback0
D        1.0.0.0/8 is a summary, 00:13:23, Null0
         12.0.0.0/8 is variably subnetted, 2 subnets, 2 masks
C        12.1.1.0/24 is directly connected, Serial1/0
D        12.0.0.0/8 is a summary, 00:13:23, Null0
```

可见，R1 上没有学习到 R2 上的 2.0.0.0 网络的路由。

在 R1 上正确配置验证之后，则会弹出 "*Nov 5 00:10:49.347: %DUAL-5-NBRCHANGE: IP-EIGRP(0) 1: Neighbor 12.1.1.1 (Serial1/0) is up: new adjacency"，说明 R1 和 R2 的验证通过，邻居关系建立起来了。

此时再查看 R1 上的路由表：

```
R1#show ip route
……
         1.0.0.0/8 is variably subnetted, 2 subnets, 2 masks
C        1.1.1.0/24 is directly connected, Loopback0
D        1.0.0.0/8 is a summary, 00:14:29, Null0
D        2.0.0.0/8 [90/2297856] via 12.1.1.2, 00:00:16, Serial1/0
         12.0.0.0/8 is variably subnetted, 2 subnets, 2 masks
C        12.1.1.0/24 is directly connected, Serial1/0
D        12.0.0.0/8 is a summary, 00:14:29, Null0
```

可见，R1 和 R2 的路由学习已正常。

实验 15 OSPF 的配置

一、实验要求

- 掌握 OSPF 的基本配置方法；
- 理解 OSPF 邻居表的基本内容；
- 理解 OSPF 拓扑表的基本内容。

二、实验说明

本实验是 OSPF 的基本配置，主要是掌握 OSPF 的一些常规配置，理解 OSPF 的邻居表和拓扑表，为后面进一步深入学习 OSPF 路由协议和工程中的应用打下基础。

本实验需要在 DynamipsGUI 来完成。

三、实验拓扑

本实验的拓扑图如图 15-1 所示。

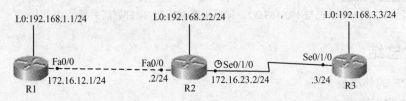

图 15-1 OSPF 的配置拓扑图

四、实验配置

1. R1 的配置过程

```
R1(config)#int f0/0
R1(config-if)#ip add 172.16.12.1 255.255.255.0        //基本配置
R1(config-if)#no shut
R1(config-if)#int L0
R1(config-if)#ip add 192.168.1.1 255.255.255.0
R1(config-if)#exit
R1(config)#router ospf 1                              //配置 OSPF 协议，进程 ID 为 1（取值范围
                                                      //1~65535），其值只有本地意义，不同路由器的进
                                                      //程 ID 互不影响
R1(config-router)#netw 192.168.1.0 0.0.0.255 area 0   //公告网络
R1(config- router)#netw 172.16.12.0 0.0.0.255 area 0
```

公告网络时，标准格式是："network 网络号 反掩码 区域号"，与 EIGRP 公告网络类似，但多了一个区域号。OSPF 路由器要形成邻接关系，需要配置相同的区域号，区域号是基于端口的，不是针对整个路由器的，一个路由器的不同端口，可以配置在不同的区域中。对单区域的 OSPF 配置，所使用区域号只能是 0。

2. R2 的配置过程

  R2(config)#router ospf 1　　　　　　　　　　//进程 ID 可以与其他路由器不相同
  R2(config-router)#netw 192.168.2.0 0.0.0.255 area 0
  R2(config-router)#netw 172.16.12.0 0.0.0.255 area 0
  R2(config-router)#netw 172.16.23.0 0.0.0.255 area 0
  04:53:51: %OSPF-5-ADJCHG: Process 1, Nbr 192.168.1.1 on FastEthernet0/0 from LOADING to FULL, Loading Done
    //提示 R2 的 f0/0 端口与 192.168.1.1，即路由器 R1 已建立好了邻居关系

从拓扑图可见，本路由器还应与 R3 建立邻居关系，现在还没有相应提示，这是因为 R3 还没配置完成，一旦 R3 配置完成，本路由器也将会产生类似提示信息。

3. R3 的配置过程

  R3(config)#router ospf 1
  R3(config-router)#netw 192.168.3.0 0.0.0.255 area 0
  R3(config-router)#netw 172.16.23.0 0.0.0.255 area 0
  04:56:23: %OSPF-5-ADJCHG: Process 1, Nbr 192.168.2.2 on Serial0/1/0 from LOADING to FULL, Loading Done
    //提示 R3 的 s0/1/0 已与 192.168.2.2，即路由器 R2 已建立好邻居关系

五、实验总结

1. OSPF 的邻居表

查看 R2 的邻居表：

  R2#show ip ospf neighbor

Neighbor ID	Pri	State	Dead Time	Address	Interface
192.168.1.1	1	FULL/DR	00:00:31	172.16.12.1	FastEthernet0/0
192.168.3.3	0	FULL/ -	00:00:34	172.16.23.3	Serial0/1/0

① **Neighbor ID**：指邻居路由器的 RID。

OSPF 确定 RID 的规则是：最优先的是在 OSPF 进程中配置 router-id 来指定 RID；其次是最大的逻辑接口 IP 地址作为 RID；最后是最大的活动物理接口 IP 地址作为 RID。

在这里，如果在 R3 上配置 router-id 来指定的 RID：

  R3(config)#router ospf 1
  R3(config-router)# router-id 3.3.3.3

然后清空进程：R3#clear ip ospf process，或重新启动路由器 R3，使配置生效。
此时再到 R2 上查看邻居表：

R2#show ip ospf neighbor

Neighbor ID	Pri	State	Dead Time	Address	Interface
192.168.1.1	1	FULL/DR	00:00:32	172.16.12.1	FastEthernet0/0
3.3.3.3	0	FULL/ -	00:00:37	172.16.23.3	Serial0/1/0

可见，邻居路由器 R3 的 RID 被修改了。

② **Pri**：邻居接口的优先级。以太网接口需要选举 DR/BDR，其默认优先级值为 1（如 R1 与 R2 间是以太网接口相连）；而点对点链路不需要选举 DR/BDR，其默认优先级值为 0（如 R2 与 R3 间是以点对点链路相连）。

③ **State**：邻居路由器的状态。"FULL/DR"中，"FULL"表示已建立了邻居关系，"DR"是指邻居被选为 DR（关于 DR/BDR 在下一个实验中详细讲述）；"FULL/ - "表示建立了邻居关系，但由于是点对点链路，不需要选择 DR 和 BDR。

④ **Dead Time**：指邻居路由器的死亡时间，当该值减少到 0 时邻居消失。

⑤ **Address**：邻居与本路由器直连的接口所配置的 IP。

⑥ **Interface**：本路由器与邻居相连的接口名。

2．OSPF 的链路状态数据库

查看 R2 的链路状态数据库 LSDB，即拓扑表：

R2#show ip ospf database
　　　　　　OSPF Router with ID (192.168.2.2) (Process ID 1)

　　　　　　Router Link States (Area 0)　　　　　　//这是路由器 LSA

Link ID	ADV Router	Age	Seq#	Checksum	Link count
192.168.3.3	192.168.3.3	1342	0x8000000a	0x00b293	3
192.168.1.1	192.168.1.1	1333	0x80000005	0x0095d5	2
192.168.2.2	192.168.2.2	1267	0x8000000f	0x007b89	4
3.3.3.3	3.3.3.3	1267	0x80000003	0x00ff14	3

　　　　　　Net Link States (Area 0)　　　　　　//这是网络 LSA

Link ID	ADV Router	Age	Seq#	Checksum
172.16.12.1	192.168.1.1	1333	0x80000003	0x000845

☞说明：

> 由于 OSPF 的链路状态数据库涉及的内容很多，在此仅对本实验中产生的链路状态数据库作简单介绍，更多的相关内容请读者参见 Cisco CCNP 相关教材。

链路状态数据库 LSAB 由链路状态通告 LSA 产生，在本实验中，产生了两种 LSA：LSA1（路由器 LSA）和 LSA2（网络 LSA）。

路由器 LSA（Router Link States (Area 0)），即 LSA1 的作用是：通告路由器所有接口的链路状态信息，包括有哪些直连网络，通过这些网络与哪些路由器相连及这些网络的代价，所有运行 OSPF 的路由器都会产生 LSA1，LSA1 只会在生成它的区域内传播，不会广播至其他的区域。

Link ID 和 **ADV Router** 是相同的，即路由器的 LSA 的 Link ID 与通告路由器的 ID 值相同。

Age 是生存时间，以秒为单位，表示 LSA 从产生到现在的时间，最长时间为 3600 秒，刷新时间为 1800 秒，如果 LSA 达到 3600，则必须从 LSDB 中清除出去。

Seq# 是序列号，当 LSA 每次产生一个新的实例，这个序列号增加 1，用于在路由器收到同一个 LSA 的多个实例时，识别最新的实例。

Checksum 是校验和，是对 LSA 的校验。

Link count 是链路数量，标明一个 LSA 所描述的路由器的链路数量，这里的链路数量与运行 OSPF 接口概念不同。

网络 LSA（Net Link States(Area 0)），即 LSA2 的作用是：LSA2 由 DR 生成，LSA2 用来描述 DR 所在的这个多路访问网络以及所有与这个多路访问网络相连的路由器 ID，包括 DR 自己。DR 将 LSA2 发送给所有的邻居，LSA2 也只在本区域内泛洪，不会被转发到区域外。

Link ID 和 **ADV Router**，在 LSA2 中，Link ID 表示 DR 的接口 IP，ADV Router 表示 DR 的 RID，这一点与 LSA1 不同。

将 LSDB 称为拓扑表，是沿用了 EIGRP 的习惯，准确的名称是链路状态数据库，里面存放的是一张整个区域内部的拓扑结构，OSPF 的每个路由器，使用 SPF 算法通过对 LSDB 的计算生成一张最短路径树（SPF 树），基于 SPF 树，计算到每一个目的网络的最佳路径，形成路由表。

3．路由表

关于路由表的相关内容，在下一个实验"OSPF 的广播多路访问"中详细讲述。

实验 16 OSPF 的广播多路访问

一、实验要求

- 了解路由器 RID 的确定顺序；
- 掌握 DR/BDR 的选择原则。

二、实验说明

OSPF 协议可以运行在三种拓扑结构中：广播型多路访问（Broadcast Multi-Access）、点到点拓扑结构、非广播型多路访问（Non-Broadcast Multi-Access，NBMA），其中广播多路访问具有将一个物理消息发送给所有在同一个网络上的路由器的能力，如以太网。在广播型多路访问和非广播型多路访问中，OSPF 必须给每个网段选举出指定路由器 DR 和备份指定路由器。其作用是：

① 减少路由更新数据流。在多路访问网络环境中，多台路由器可以互为邻居，如果它们之间都建立相邻关系并相互交换链路状态信息，则关系复杂，数据流量大。在选举了 BR 和 DBR 之后，每台路由器都只与 DR 和 BDR 建立相邻关系和交换链路状态信息，这种扩散过程大大减小了网络上的数据流量。

② 管理链路状态同步。DR 和 BDR 可以保证网络上的其他路由器的链路状态信息是同步的。

本实验主要学习 OSPF 在广播多路访问环境下的配置，RID 的确认方法，DR/BDR 的选举原则，端口优先级的更改以及如何改变 DR/DBR 的选举。

三、实验拓扑

本实验拓扑如图 16-1 所示。

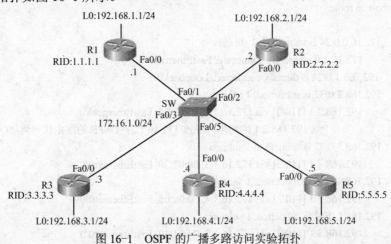

图 16-1 OSPF 的广播多路访问实验拓扑

四、配置过程

由于各路由器的配置过程是相似的，在此只配置路由器 R1，其余路由器的配置可按拓扑标示，参照 R1 进行配置即可。

1. 交换机 SW 的配置过程

  Switch>en
  Switch#conf t
  Switch(config)#host SW
  SW(config)#no cdp run   //关闭 CDP 协议，

2. R1 的配置过程

  Router>en
  Router#conf t
  Router(config)#host R1
  R1(config)#int L0
  R1(config-if)#ip add 192.168.1.1 255.255.255.0
  R1(config-if)#int f0/0
  R1(config-if)#ip add 172.16.1.1 255.255.255.0
  R1(config-if)#no shut
  R1(config-if)#exit
  R1(config)#router ospf 1   //启用 OSPF 进程，进程号为 1（范围是 1~65535）
  R1(config-router)#router-id 1.1.1.1   //配置 RID 为 1.1.1.1
  R1(config-router)#netw 192.168.1.0 0.0.0.255 area 0   //公告直连网络
  R1(config-router)#netw 172.16.1.0 0.0.0.255 area 0

将路由器 R2、R3、R4、R5 按类似的方法配置完成。

五、实验总结

1. 查看路由表

```
R1#show ip route
……
     172.16.0.0/24 is subnetted, 1 subnets
C       172.16.1.0 is directly connected, FastEthernet0/0
C    192.168.1.0/24 is directly connected, Loopback0
     192.168.2.0/32 is subnetted, 1 subnets
O       192.168.2.1 [110/2] via 172.16.1.2, 00:02:25, FastEthernet0/0
                //到 192.168.2.1 的路由，经过 172.16.1.2；OSPF 的路由代码为 "O"
     192.168.3.0/32 is subnetted, 1 subnets
O       192.168.3.1 [110/2] via 172.16.1.3, 00:07:30, FastEthernet0/0
     192.168.4.0/32 is subnetted, 1 subnets
O       192.168.4.1 [110/2] via 172.16.1.4, 00:06:05, FastEthernet0/0
     192.168.5.0/32 is subnetted, 1 subnets
O       192.168.5.1 [110/2] via 172.16.1.5, 00:04:30, FastEthernet0/0
```

可见，已经正常学习到全网的 OSPF 路由条目，这是继续查看其他性质的基础。

2. 路由器 RID 的确定顺序

首先是在 OSPF 进程中通过配置 router-id 指定 RID。在本实验中，每个路由器都通过 router-id 指定了 RID。

如果没有配置 router-id 指定的 RID，则在所配置的环回接口中选最大 IP 地址作为 RID。

如果没有配置环回接口，则选择最大活动物理接口的 IP 地址为 RID。

3. DR/BDR 的选举

DR 和 BDR 的选举规则如下。

首先比较的是接口优先级，优先级最高的路由器为 DR，次高的为 BDR。凡是参与了 DR/BDR 选举的接口，其优先级默认都是"1"。

如接口优先级相同，RID 最高的为 DR，次高的为 BDR，其余的为 DROTHER。

要查看路由器的 DR/BDR 选举情况，可通过 "show ip ospf neighbor" 来查看 OSPF 邻居信息：

R1#show ip ospf neighbor

Neighbor ID	Pri	State	Dead Time	Address	Interface
3.3.3.3	1	FULL/BDR	00:00:37	172.16.1.3	FastEthernet0/0
4.4.4.4	1	FULL/DROTHER	00:00:39	172.16.1.4	FastEthernet0/0
5.5.5.5	1	FULL/DROTHER	00:00:36	172.16.1.5	FastEthernet0/0
2.2.2.2	1	FULL/DROTHER	00:00:39	172.16.1.2	FastEthernet0/0

从 R1 上查看 OSPF 邻居的输出，可以看到，DBR 的 Neighbor ID（邻居 ID，也就是 RID）为 3.3.3.3，而邻居 ID 为 2.2.2.2，4.4.4.4，5.5.5.5 的路由器均为 DROTHER。

再查看路由器 R2 的 OSPF 邻居：

R2#show ip ospf neighbor

Neighbor ID	Pri	State	Dead Time	Address	Interface
1.1.1.1	1	FULL/DR	00:00:37	172.16.1.1	FastEthernet0/0
5.5.5.5	1	2WAY/DROTHER	00:00:31	172.16.1.5	FastEthernet0/0
3.3.3.3	1	FULL/BDR	00:00:33	172.16.1.3	FastEthernet0/0
4.4.4.4	1	2WAY/DROTHER	00:00:34	172.16.1.4	FastEthernet0/0

由 R2 的 OSPF 邻居输出可见，DR 的邻居 ID 为 1.1.1.1，即路由器 R1。

这与前面所说的选举规则相矛盾：所有的接口 Pri（优先级）在前面的配置中，均没修改过，默认为 1；选举的 DR 应该由 RID 最大（5.5.5.5）的路由器，即 R5 来充当。

这是因为前面所述的 DR/BDR 选举规则，是在参与选举的路由器同时启动的情况下设定的。为了保证网络稳定性，在启动时，如果先启动 R1，则 R1 将成为 DR，其次启动的成为 BDR，后面启动的成为 DROther。即使后面启动的路由器配置更高，在没有管理员干预的情况下，也需要等到前面的 DR 或 BDR 离开网络后才能被选为 DR 或 BDR，并且，在前面 DR 离开后，BDR 自动升为 DR，然后从余下路由器中选优先级或 RID 最高的作为 BDR。

在此作如下验证。

在 R1 上，关闭 f0/0，造成 R1 与 OSPF 失去联系，使 R1 离开：

    R1(config)#int f0/0
    R1(config-if)#shut

当 OSPF 收敛稳定后，再查看 R2 的 OSPF 邻居：

    R2#show ip ospf neighbor

Neighbor ID	Pri	State	Dead Time	Address	Interface
5.5.5.5	1	FULL/BDR	00:00:38	172.16.1.5	FastEthernet0/0
3.3.3.3	1	FULL/DR	00:00:39	172.16.1.3	FastEthernet0/0
4.4.4.4	1	2WAY/DROTHER	00:00:31	172.16.1.4	FastEthernet0/0

可见，此时 DR 为原来的 BDR，即 R3；以前 RID 最高的 R5，现升为 BDR。

如果此时让 R1 回到 OSPF 中：

    R1(config)#int f0/0
    R1(config-if)#no shut

当 OSPF 收敛稳定后，再查看 R2 的 OSPF 邻居：

    R2#show ip ospf neighbor

Neighbor ID	Pri	State	Dead Time	Address	Interface
5.5.5.5	1	FULL/BDR	00:00:37	172.16.1.5	FastEthernet0/0
3.3.3.3	1	FULL/DR	00:00:38	172.16.1.3	FastEthernet0/0
4.4.4.4	1	2WAY/DROTHER	00:00:30	172.16.1.4	FastEthernet0/0
1.1.1.1	1	2WAY/DROTHER	00:00:33	172.16.1.1	FastEthernet0/0

可见，DR 仍是 R3，BDR 仍是 R5，R1 没能"夺回"DR 的位置，一方面是由于 DR 和 BDR 工作正常，另一方面是 R1 的 RID 低。

4. 修改端口优先级

在广播多路访问中，为使性能更好的路由器来充当 DR、BDR，可以通过修改端口优先级的方法，来改变 DR/BDR 的选举。

把性能好的路由器，其参与 OSPF 进程中 DR/BDR 选举的端口改成大于 1 的数值；而将性能差的，永远不使之成为 DR/BDR 的路由器端口优先级改为 0，这样该端口就不参与 DR/BDR 的选举了。修改的配置命令如下：

    R1(config)#int f0/0
    R1(config-if)#ip ospf priority 10
           //将 R1 的 f0/0 端口优先级改为 10
    R1(config-if)#end
    R1#clear ip ospf process
           //除了重启端口外，还可以重启 OSPF 进程让 DR/BDR 重新选举
    R3(config)#int f0/0
    R3(config-if)#ip ospf priority 0
           //将 R3 的 f0/0 端口优先级改为 0，让此端口永远不参与 DR/BDR 的选举
    R3(config-if)#end
    R3#clear ip ospf process

实验 17 OSPF 默认路由

一、实验要求

- 了解注入默认路由的作用；
- 掌握向 OSPF 网络中注入默认路由的方法。

二、实验说明

在前面已经讲过 RIP 的默认路由配置，本实验重点讲述在 OSPF 路由协议中注入默认路由，在图 17-1 中，R1 是企业边界路由器，内网配置了 OSPF 路由协议，通过在 R1 上的宣告一条动态的默认路由，在内部路由器 R2 和 R3 上，就不需要配置去往边界路由器的默认路由了，从而减少了内网配置的复杂性。

本实验在 DynamipsGUI 环境下完成。

三、实验拓扑

实验拓扑如图 17-1 所示。

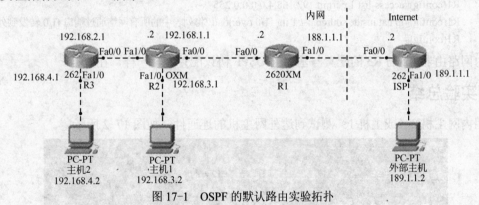

图 17-1 OSPF 的默认路由实验拓扑

四、实验过程

（1）ISP 路由器的配置

  Router#conf t
  R3(config)#host ISP
  ISP(config)#int f0/0
  ISP(config-if)#ip add 188.1.1.2 255.255.255.0
  ISP(config-if)#int f1/0
  ISP(config-if)#ip add 189.1.1.1 255.255.255.0
  ISP(config-if)#no shut

（2）边界路由器 R1 的配置

  Router#conf t

```
Router(config)#host R1
R1(config)#int f0/0
R1(config-if)#ip add 192.168.1.2 255.255.255.0       //基本配置
R1(config-if)#no shut
R1(config-if)#int f1/0
R1(config-if)#ip add 188.1.1.1 255.255.255.0
R1(config-if)#no shut
R1(config-if)#exit
R1(config)#router ospf 1
R1(config-router)#netw 192.168.1.0 0.0.0.255 area 0  //只公告内部网络，不能公告与外网相连的网络
R1(config-router)#default-information originate       //声明 R1 是默认路由起源，R1 则会向其
R1(config-router)#exit                                //他 RIP 路由器宣告自己是默认路由
R1(config)#ip route 0.0.0.0 0.0.0.0 188.1.1.2         //配置默认路由，实现对 Internet 的访问
R1(config)#int f0 0
R1(config-if)#ip nat inside                           // 指定 PAT 地址转换的内外端口(关于网
R1(config-if)#int f1/0                                //络地址转换，后面章节会专门讲述)
R1(config-if)#ip nat outside
R1(config-if)#exit
R1(config)#access-list 1 permit 192.168.1.0 0.0.0.255 //定义访问控制列表
R1(config)#access-list 1 permit 192.168.2.0 0.0.0.255
R1(config)#access-list 1 permit 192.168.3.0 0.0.0.255
R1(config)#access-list 1 permit 192.168.4.0 0.0.0.255
R1(config)#ip nat inside source ;ist 1 int f1/0 overload//将列表中的所有网络通过端口 f1/0 转发到外网
R1(config)#
```

内网路由器 R2 与 R3 的配置与基本的 OSPF 协议配置完全一样，在此略。

五、实验总结

用内网主机 2（或主机 1）测试到达外网主机的连通性，如图 17-2 所示。

```
PC>ping 189.1.1.2

Pinging 189.1.1.2 with 32 bytes of data:

Request timed out.
Request timed out.
Reply from 189.1.1.2: bytes=32 time=18ms TTL=125
Reply from 189.1.1.2: bytes=32 time=16ms TTL=125

Ping statistics for 189.1.1.2:
    Packets: Sent = 4, Received = 2, Lost = 2 (50% loss),
Approximate round trip times in milli-seconds:
    Minimum = 16ms, Maximum = 18ms, Average = 17ms
```

图 17-2　测试到达外网主机的连通性

可见，从内网主机能访问外网主机。

然后去查看边界路由器 R1 的路由表：

```
R1#show ip route
……
```

```
            188.1.0.0/24 is subnetted, 1 subnets
   C        188.1.1.0 is directly connected, FastEthernet1/0
   C        192.168.1.0/24 is directly connected, FastEthernet0/0
   O        192.168.2.0/24 [110/2] via 192.168.1.1, 00:09:08, FastEthernet0/0
   O        192.168.3.0/24 [110/2] via 192.168.1.1, 00:08:58, FastEthernet0/0
   O        192.168.4.0/24 [110/3] via 192.168.1.1, 00:07:51, FastEthernet0/0
   S*       0.0.0.0/0 [1/0] via 188.1.1.2
```

可见，与前面 RIP 一样，在 R1 上有一条默认路由到 ISP 的入口，它的作用是将内网中转发来的所有未知网络，转到 ISP 的接入路由器。

再查看内网路由器 R2 的路由表：

```
R2#show ip route
……
   C        192.168.1.0/24 is directly connected, FastEthernet0/0
   C        192.168.2.0/24 is directly connected, FastEthernet1/1
   C        192.168.3.0/24 is directly connected, FastEthernet1/0
   O        192.168.4.0/24 [110/2] via 192.168.2.1, 00:10:23, FastEthernet1/1
   O*E2 0.0.0.0/0 [110/1] via 192.168.1.2, 00:11:41, FastEthernet0/0
```

可见，R3 学到一条"O*E2 0.0.0.0/0 [110/1] via 192.168.1.2, 00:11:41, FastEthernet0/0"路由，"O*"表示是通过 OSPF 学来的默认路由；"E2"表示该路由是 OSPF 外部类型 2 的路由。OSPF 有两种外部类型路由，用"E1"和"E2"来表示，它们的区别是 1 型外部路由既要算从外部进来的花费，还要算上 OSPF 域内的花费；而 2 型外部路由只算外部进来的花费。因此，1 型外部路由更准确。

如果关闭边界路由器 R1 与 ISP 相连的接口 f1/0：

```
R1(config)#int f1/0
R1(config-if)#shut
```

然后再去查看 R2（或 R3）的路由表：

```
R2#show ip route
……
   C        192.168.1.0/24 is directly connected, FastEthernet0/0
   C        192.168.2.0/24 is directly connected, FastEthernet1/1
   C        192.168.3.0/24 is directly connected, FastEthernet1/0
   O        192.168.4.0/24 [110/2] via 192.168.2.1, 00:19:43, FastEthernet1/1
```

可见，R2（或 R3）的路由表中默认路由消失了，这是因为只有当 R1 上有向外发布的默认路由时，"default-information originate"才会使 R1 向外发布默认路由有效。现在 R1 的外出接口被 shutdown 了，也就使 R1 不再向外发布默认路由，R1 也就不能再向内网路由器 R2 和 R3 注入默认路由了。

在边界路由器 R1 上，配置向外的发布默认路由命令时，如果加上"always"参数，即"default-information originate always"，可使内部路由器的那条 OSPF 默认路由不再消失（注：加了"always"的"default-information originate always"命令在 Cisco Packet Tracer 中不支持）。

实验 18 OSPF 验证

一、实验要求

- 掌握 OSPF 的 text 验证的配置方法；
- 掌握 OSPF 的 MD5 验证的配置方法。

二、实验说明

OSPF 支持明文的 text 和密文的 MD5 两种验证，通过验证功能，能保障路由器之间的可靠更新。

text 验证将会在网上以明文方式发送验证密码，而 MD5 模式则采用密文方式，安全性比 text 模式更高（本实验需要使用 DynamipsGUI 来完成，因为 Cisco Packet Tracer 不支持 MD5 验证）。

其配置过程如下：
① 定义钥匙链；
② 在钥匙链上定义一个或者一组钥匙；
③ 在接口上启用认证并指定使用的钥匙链。

三、实验拓扑

实验拓扑如图 18-1 所示。

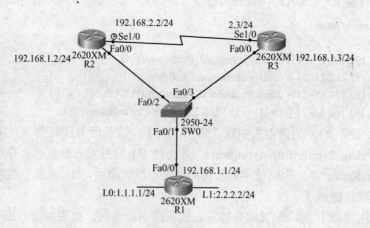

图 18-1 OSPF 验证实验拓扑

四、实验过程及实验总结

根据图 18-1 所示，完成 OSPF 的基本配置。

然后，在 R1 上查看当前的验证方式：

```
R1#debug ip ospf packet                              //查看 OSPF 分组接收情况
OSPF packet debugging is on                          //提示 debug 已开启
R1#
*Dec 24 23:22:34.875: OSPF: rcv. v:2 t:1 l:52 rid:192.168.2.2
       aid:0.0.0.0 chk:DFE5 aut:0 auk: from FastEthernet0/0    //aut:0 中的"0"表示无须验证，如果
                                                     //是"1"，表示明文验证方式；如果
                                                     //是"2"，表示 MD5 验证方式
R1#no debug all                                      //关闭 debug，因为开启 debug 会占
                                                     //用系统资源
All possible debugging has been turned off           //提示 debug 已关闭
```

验证分为针对区域和针对链路的验证，下面分别讲述明文验证和 MD5 验证针对区域的验证方法。

1. 明文验证

在 R1 上配置明文验证：

```
R1(config)#int f0/0
R1(config-if)#ip ospf authentication-key abc         //在接口 f0/0 上配置明文密码 abc
R1(config-if)#router ospf 1
R1(config-router)#area 0 authentication              //声明 area 0 使用明文验证
R1(config-router)#
   *Dec 24 23:34:54.851: %OSPF-5-ADJCHG: Process 1, Nbr 192.168.2.2 on FastEthernet/0 from
FULL to DOWN, Neighbor Down: Dead timer expired
   *Dec 24 23:34:56.447: %OSPF-5-ADJCHG: Process 1, Nbr 192.168.2.3 on FastEthernet0/0 from
FULL to DOWN, Neighbor Down: Dead timer expired
```

最后这几行是提示，R1 的两个邻居 192.168.2.2 和 192.168.2.3 已经断开（down）了，这是因为只在 R1 上启用了验证，而 R2 和 R3 上没有启用验证，在前面讲 OSPF 包类型中的 hello 包时讲过，如果有验证，需要有相同的验证才能相互发送 hello 分组，从而建立起邻居关系（可以分别在 R1 和 R2 上，启用"debug ip ospf packet"命令查看验证方式，它们是不相同的，R1 上已经没有收到 OSPF 分组的相关信息了，而 R2 是 0）。

然后，分别在 R2 和 R3 上启用明文验证：

```
R2(config)#int f0/0
R2(config-if)#ip ospf authentication-key abc         //整个区域都需要配置相同的密码，下同
R2(config-if)#int s1/0
R2(config-if)#ip ospf authentication-key abc
R2(config-if)#router ospf 100
R2(config-router)#area 0 authentication

R3(config)#int f0/0
R3(config-if)#ip ospf authentication-key abc
R3(config-if)#int s1/0
R3(config-if)#ip ospf authentication-key abc
R3(config-if)#router ospf 1
```

```
R3(config-router)#area 0 authentication
```

配置完成后，可见到邻居关系建立成功的提示，邻接关系恢复正常。

明文验证的缺点是可以用协议分析仪在网上窃取到密码，从而容易受到攻击。而下面讲述的 MD5 则是非常安全的验证方式。

2．MD5 验证

MD5 验证算法的安全性比明文验证高，并且采用 MD5 验证，OSPF 每个接口同时支持两套密钥，如果配置的密钥不小心泄露了，那么可以在路由器的相应接口上再配置一套密钥，并且把旧的密钥删除掉。在配置新密钥的过程中，OSPF 路由器将向网络发送同一个报文的两份拷贝，可分别用新旧密钥来验证，这样就不会中断网络正常通信功能。

MD5 验证的配置如下：

```
R1(config)#int f0/0
R1(config-if)#ip ospf message-digest-key 1 md5 xyz    //在 f0/0 上启用 MD5，密码 ID 为 1，密
                                                      //码为 xyz
R1(config-if)#router ospf 1
R1(config-router)#area 0 authentication message-digest   //在区域 0 上使用 MD5 验证

R2(config)#int f0/0
R2(config-if)#ip ospf message-digest-key 1 md5 xyz    //整个区域都需要配置相同的密码，下同
R2(config-if)#int s1/0
R2(config-if)#ip ospf message-digest-key 1 md5 xyz
R2(config-if)#router ospf 100
R2(config-router)#area 0 authentication message-digest

R3(config)#int f0/0
R3(config-if)#ip ospf message-digest-key 1 md5 xyz
R3(config-if)#int s1/0
R3(config-if)#ip ospf message-digest-key 1 md5 xyz
R3(config-if)#router ospf 1
R3(config-router)#area 0 authentication message-digest
```

配置完成后，所有路由器邻居关系重新建立，网络恢复正常。（读者可以在 R1 的 f0/0 接口上再配置一个 MD5 密钥，看是否会造成 R1 与 R2 和 R3 间的邻接关系中断。）

实验 19 多区域 OSPF 的配置

一、实验要求

- 掌握 OSPF 的基本配置方法；
- 掌握多区域 OSPF 的配置方法；
- 掌握 RIP 与 OSPF 的路由重发布；
- 在 OSPF 区域间实现路由归纳。

二、实验说明

多区域 OSPF 中涉及以下两个重要概念。

区域边界路由器（Area Border Router，ABR）：位于一个或多个 OSPF 区域边界上，将这些区域连接到主干网络的路由器，如图 19-1 的路由器 R2。

自治系统边界路由器（Autonomous System Border Router，ASBR）：ASBR 位于 OSPF 自主系统和非 OSPF 网络之间。ASBR 可以运行 OSPF 和另一路由选择协议（如 RIP），将其他路由协议重发布到 OSPF 上，如图 19-1 所示的路由器 R1。

在图 19-1 中，R1 是自治系统边界路由器，在 R1 上同时运行两个路由协议进程：RIP 及 OSPF，它通过 RIP 进程学习到 RIP 路由，又通过 OSPF 进程学习到 OSPF 域内的路由，RIP 和 OPSF 是独立的，默认情况下，路由信息不会相互传递，这就需要实现 RIP 路由和 OSPF 路由的相互注入，这就是路由的重发布。

本实验在 DynamipsGUI 中完成。

三、实验拓扑

实验采用的拓扑如图 19-1 所示。

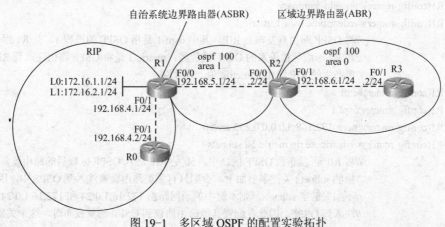

图 19-1 多区域 OSPF 的配置实验拓扑

四、实验过程及实验总结

1. OSPF 的重发布

（1）R0 的配置过程

```
Router#conf t
Router(config)#host R0
R0(config)#int f0/1
R0(config-if)#ip address 192.168.4.2 255.255.255.0
R0(config-if)#no shut
R0(config-if)#exit
R0(config)#router rip
R0(config-router)#version 2
R0(config-router)# network 192.168.4.0
R0(config-router)# no auto-summary
R0(config-router)#end
R0#
```

（2）R1 的配置过程

```
Router>en
Router#conf t
Router(config)#host R1
R1(config)#int L0
R1(config-if)#ip add 172.16.1.1 255.255.255.0
R1(config-if)#int L1
R1(config-if)#ip add 172.16.2.1 255.255.255.0
R1(config-if)#int f0/0
R1(config-if)#ip add 192.168.5.1 255.255.255.0
R1(config-if)#no shut
R1(config-if)#exit
R1(config)#router rip
R1(config-router)#ver 2
R1(config-router)#netw 172.16.0.0
R1(config-router)#no auto-summary
R1(config-router)#redistribute ospf 1 metric 3
          //将 OSPF 路由重发布到 RIP，其中 ospf 1 是指 OSPF 的进程 1，是 R1 用于
          //和 R2 形成邻接关系的 OSPF 进程号；metric 3 是将 OSPF 路由注入到 RIP
          //所产生的 RIP 路由的 metric 值，即 3 跳
R1(config-router)#exit
R1(config)#router ospf 1
R1(config-router)#netw 192.168.5.0 0.0.0.255 area 0
R1(config-router)#redistribute rip metric 50 subnets
          //将 RIP 再发布到 OSPF 区域中，50 是 RIP 进入 OSPF 区域后的路由度量值。这
          //里的 subnets 关键字要加上，否则只有主类路由会被注入到 OSPF 中，因此如果
          //不加关键字 subnets，则本例中的子网路由 172.16.1.0/24 和 172.16.1.0/24 就无法
          //注入到 OSPF。因此在配置其他路由协议到 OSPF 的重发布时，这个关键字一
```

```
                        //一般都是要加上的
        R1(config-router)#router-id 1.1.1.1
        R1(config-router)#end
        R1#
```

（3）R2 的配置过程

```
        Router>en
        Router#conf t
        Router(config)#host R2
        R2(config)#int f0/1
        R2(config-if)#ip add 192.168.6.1 255.255.255.0
        R2(config-if)#no shut
        R2(config-if)#int f0/0
        R2(config-if)#ip add 192.168.5.2 255.255.255.0
        R2(config-if)#no shut
        R2(config-if)#exit
        R2(config)#router ospf 100
        R2(config-router)#router-id 2.2.2.2
        R2(config-router)#netw 192.168.5.0 0.0.0.255 area 1
        R2(config-router)#netw 192.168.6.0 0.0.0.255 area 0
```

（4）R3 的配置过程

```
        Router>en
        Router#conf t
        Router(config)#host R3
        R3(config)#int f0/1
        R3(config-if)#ip add 192.168.6.2 255.255.255.0
        R3(config-if)#no shut
        R3(config-if)#router ospf 100
        R3(config-router)#router-id 3.3.3.3
        R3(config-router)#network 192.168.6.0 0.0.0.255 area 0
```

配置完成后，查看各路由器的路由表。

（5）R0 的路由表

```
        R0#show ip route
        ……
                172.16.0.0/24 is subnetted, 2 subnets
        R          172.16.1.0 [120/1] via 192.168.4.1, 00:00:25, FastEthernet0/1
        R          172.16.2.0 [120/1] via 192.168.4.1, 00:00:25, FastEthernet0/1
        C          192.168.4.0/24 is directly connected, FastEthernet0/1
        R          192.168.5.0/24 [120/3] via 192.168.4.1, 01:10:06, FastEthernet0/1
        R          192.168.6.0/24 [120/3] via 192.168.4.1, 01:10:06, FastEthernet0/1
```

最后两条到达 192.168.5.0/24 和 192.168.6.0/24 的路由，是从 OSPF 域中重发布进来的。

（6）R1 的路由表

```
R1#show ip route
……
      172.16.0.0/24 is subnetted, 2 subnets
C        172.16.1.0 is directly connected, Loopback0
C        172.16.2.0 is directly connected, Loopback1
C     192.168.4.0/24 is directly connected, FastEthernet0/1
C     192.168.5.0/24 is directly connected, FastEthernet0/0
O IA  192.168.6.0/24 [110/2] via 192.168.5.2, 00:01:04, FastEthernet0/0
```

最后一条路由，前面带有"O IA"标记的，表示是一条区域间的外部路由，是从 OSPF 的 area 0 发过来的。

（7）R2 的路由表

```
R2#show ip route
……
      172.16.0.0/24 is subnetted, 2 subnets
O E2     172.16.1.0 [110/50] via 192.168.5.1, 00:04:31, FastEthernet0/0
O E2     172.16.2.0 [110/50] via 192.168.5.1, 00:04:31, FastEthernet0/0
O E2  192.168.4.0/24 [110/50] via 192.168.5.1, 01:18:31, FastEthernet0/0
C     192.168.5.0/24 is directly connected, FastEthernet0/0
C     192.168.6.0/24 is directly connected, FastEthernet0/1
```

其中前面带有"O E2"标记的，表示外部路由，是从非 OSPF 区域（这里是 RIP）发送进来的。

（8）R3 的路由表

```
R3#show ip route
Codes: C - connected, S - static, I - IGRP, R - RIP, M - mobile, B - BGP
       D - EIGRP, EX - EIGRP external, O - OSPF, IA - OSPF inter area
       N1 - OSPF NSSA external type 1, N2 - OSPF NSSA external type 2
       E1 - OSPF external type 1, E2 - OSPF external type 2, E - EGP
       i - IS-IS, L1 - IS-IS level-1, L2 - IS-IS level-2, ia - IS-IS inter area
       * - candidate default, U - per-user static route, o - ODR
       P - periodic downloaded static route

Gateway of last resort is not set

      172.16.0.0/24 is subnetted, 2 subnets
O E2     172.16.1.0 [110/50] via 192.168.6.1, 01:09:04, FastEthernet0/1
O E2     172.16.2.0 [110/50] via 192.168.6.1, 01:09:04, FastEthernet0/1
O E2  192.168.4.0/24 [110/50] via 192.168.6.1, 01:12:56, FastEthernet0/1
O IA  192.168.5.0/24 [110/2] via 192.168.6.1, 01:12:56, FastEthernet0/1
C     192.168.6.0/24 is directly connected, FastEthernet0/1
```

前面带有"O IA"标记的，表示是一条区域间的外部路由，是从 OSPF 的 area 1 发送到 area 0 的；前面带有"O E2"标记的，表示外部路由，是从非 OSPF 区域（这里是 RIP）发送进来的。

2. 重发布直连路由到 OSPF 区域

实验采用的拓扑如图 19-2 所示。

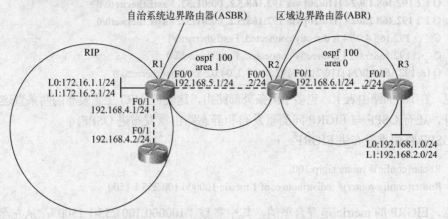

图 19-2 重发布直连路由到 OSPF 区域实验拓扑

在图 19-1 的基础上，在 R3 上新增两个环回接口，但这两个环回接口并没有公告进 OSPF 区域，这样路由器 R0、R1、R2 就不能通过 OSPF 学习这个接口的直连路由。

现在可以在 R3 上对直连路由进行重发布，将 R3 的直连路由注入到 OSPF 域中，形成 OSPF 外部路由，然后通告给域内的其他路由器。

```
R3(config)#router ospf 100
R3(config-router)#redistribute connected subnets
```

通过使用 "redistribute connected subnets" 命令后，R1 上所有未被 OSPF network 命令激活 OSPF 的接口的直连路由都会被注入到 OSPF 中。

查看 R3 的路由表：

```
R3#
%SYS-5-CONFIG_I: Configured from console by console

R3#show ip route
……
        172.16.0.0/24 is subnetted, 2 subnets
O E2    172.16.1.0 [110/50] via 192.168.6.1, 03:38:13, FastEthernet0/1
O E2    172.16.2.0 [110/50] via 192.168.6.1, 03:38:13, FastEthernet0/1
C       192.168.1.0/24 is directly connected, Loopback0
C       192.168.2.0/24 is directly connected, Loopback2
O E2 192.168.4.0/24 [110/50] via 192.168.6.1, 03:42:05, FastEthernet0/1
O IA 192.168.5.0/24 [110/2] via 192.168.6.1, 03:42:05, FastEthernet0/1
C       192.168.6.0/24 is directly connected, FastEthernet0/1
```

可见，在 R3 上多了两条直连路由，然后，再到 R1 上，查看路由表。

```
R1#show ip route
……
        172.16.0.0/24 is subnetted, 2 subnets
```

```
C       172.16.1.0 is directly connected, Loopback0
C       172.16.2.0 is directly connected, Loopback1
O E2 192.168.1.0/24 [110/20] via 192.168.5.2, 00:01:52, FastEthernet0/0
O E2 192.168.2.0/24 [110/20] via 192.168.5.2, 00:01:52, FastEthernet0/0
C       192.168.4.0/24 is directly connected, FastEthernet0/1
C       192.168.5.0/24 is directly connected, FastEthernet0/0
O IA 192.168.6.0/24 [110/2] via 192.168.5.2, 04:02:53, FastEthernet0/0
```

可见，在 R1 的路由表上，也多了两条外部路由，这就是从 R3 上重发布的两条直连路由。另外，还有 OSPF 与 EIGRP 间的重发布和静态路由重发布进 OSPF。

将 OSPF 路由重发布进 EIGRP：

```
Router(config)# router eigrp 100
Router(config-router)# redistribute ospf 1 metric 100000 100 255 1 1500
```

注意，EIGRP 的 metric 是混合型的，其中参数"100000 100 255 1 1500"，从左至右依次是"带宽、延迟、负载、可靠性、MTU"。可根据实际需要灵活地进行设定。上述配置完成后，Router 会将路由表中的 OSPF 路由和宣告进 OSPF 的直连网段注入到 EIGRP 进程。

将 EIGRP 路由重发布进 OSPF：

```
Router(config)# router ospf 100
Router(config-router)# redistribute eigrp 100 subnets
```

静态路由重发布进 OSPF：

```
Router(config)# router ospf 100
Router(config-router)# redistribute static subnets
```

在 Router 的路由表中，所有的静态路由都会被注入到 OSPF 中形成 OSPF 外部路由，并且通过 OSPF 动态地传递到整个 OSPF 域。

3．多区域的 OSPF 路由归纳

实验采用的拓扑如图 19-3 所示。

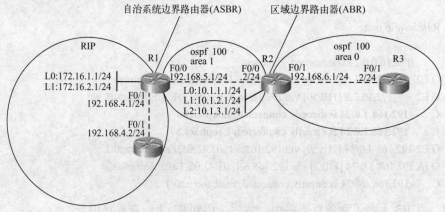

图 19-3　多区域的 OSPF 路由归纳实验拓扑

在图 19-1 的基础上，在路由器 R2 上再配置 3 个环回接口，并公告到 area 1 中，如图 19-3 所示。

（1）在 R2 上新增的配置过程：

110

```
R2(config)#int L0
R2(config-if)#ip add 10.1.1.1 255.255.255.0
R2(config-if)#int L1
R2(config-if)#ip add 10.1.2.1 255.255.255.0
R2(config-if)#int L2
R2(config-if)#ip add 10.1.3.1 255.255.255.0
R2(config-if)#exit
R2(config)#router ospf 100
R2(config-router)#netw 10.1.1.0 0.0.0.255 area 1
R2(config-router)#netw 10.1.2.0 0.0.0.255 area 1
R2(config-router)#netw 10.1.3.0 0.0.0.255 area 1
R2(config-router)#end
```

然后，在 R3 查看新的路由表：

```
R3#show ip route
……
     10.0.0.0/32 is subnetted, 3 subnets
O IA    10.1.1.1 [110/2] via 192.168.6.1, 00:00:19, FastEthernet0/1
O IA    10.1.2.1 [110/2] via 192.168.6.1, 00:00:09, FastEthernet0/1
O IA    10.1.3.1 [110/2] via 192.168.6.1, 00:00:09, FastEthernet0/1
     172.16.0.0/24 is subnetted, 2 subnets
O E2    172.16.1.0 [110/50] via 192.168.6.1, 02:52:24, FastEthernet0/1
O E2    172.16.2.0 [110/50] via 192.168.6.1, 02:52:24, FastEthernet0/1
O E2 192.168.4.0/24 [110/50] via 192.168.6.1, 02:56:15, FastEthernet0/1
O IA 192.168.5.0/24 [110/2] via 192.168.6.1, 02:56:15, FastEthernet0/1
C    192.168.6.0/24 is directly connected, FastEthernet0/1
```

可看到 R1 上的 3 条 **10.1.X.X**/24 的 OSPF 区间路由，为了减少 OSPF 区域间传递的路由条目，减小路由表的大小，可在路由器 R2 上对区域 1 的 **10.1.X.X**/24 的路由进行路由归纳：

```
R2(config)#router ospf 100
R2(config-router)#area 1 range 10.1.0.0 255.255.0.0
```

然后再次查看 R3 上的路由表：

```
R3#show ip route
……
     10.0.0.0/32 is subnetted, 1 subnets
O IA    10.1.0.0[110/2] via 192.168.6.1, 00:01:12, FastEthernet0/1
     172.16.0.0/24 is subnetted, 2 subnets
O E2    172.16.1.0 [110/50] via 192.168.6.1, 02:52:24, FastEthernet0/1
O E2    172.16.2.0 [110/50] via 192.168.6.1, 02:52:24, FastEthernet0/1
O E2 192.168.4.0/24 [110/50] via 192.168.6.1, 02:56:15, FastEthernet0/1
O IA 192.168.5.0/24 [110/2] via 192.168.6.1, 02:56:15, FastEthernet0/1
C    192.168.6.0/24 is directly connected, FastEthernet0/1
```

以前 R3 上的 3 条 10.1.X.X/24 的区间路由变成了一条汇总路由，采用多区域的 OSPF 路由归纳大大减小了主干路由器的路由表大小，最终是减少了查询路由的延时时间，提高路由转发效率。

实验 20　VLAN 基本配置

一、实验要求
- 掌握 VLAN 的基本配置方法；
- 理解 VLAN 干道的作用；
- 掌握干道许可 VLAN 的配置方法。

二、实验说明
① 同一个交换机上的端口，如果配置到同一个 VLAN，可直接通信。

② 不同交换机上的端口，如果配置到同一个 VLAN，需要将交换机之间相连的端口设置成 Trunk 模式后，才可相互通信。

③ 不同的 VLAN 间，需要配置 VLAN 间路由功能后，才能相互通信。

在本实验中主要实现的是前两项，而 VLAN 间路由在下一个实验完成。

三、实验拓扑
本实验使用的拓扑图如图 20-1 所示。

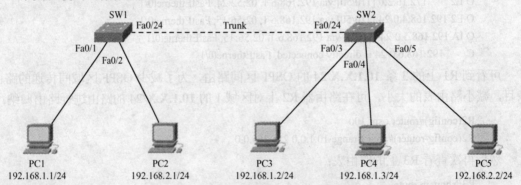

图 20-1　VLAN 基本配置实验拓扑

四、配置过程
首先按图 20-1 所示配置各 PC 的 IP 地址，以 PC1 为例，如图 20-2 所示。

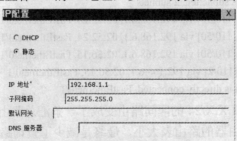

图 20-2　配置 PC1 的 IP 地址

在 PC1 的 IP 地址配置过程中，没有配置默认网关地址，如果需要实现 VLAN 间的路由功能，则需要配置默认网关地址（VLAN 间路由在下一个实验中讲述）；也没有配置 DNS 服务器地址，DNS 一般只在访问外网时才需要配置。

（1）配置交换机 SW1

```
Switch>en
Switch#conf t
Switch(config)#host SW1
SW1(config)#vlan 1                              //进入 VLAN 1
SW1(config-vlan)#name V1
Default VLAN 1 may not have its name changed.   //提示：VLAN 1 是默认 VLAN，不能改名
SW1(config-vlan)#exit
SW1(config)#vlan 2
SW1(config-vlan)#name V2                        //创建 VLAN2
                                                //将 VLAN2 重命名为 V2
SW1(config-vlan)#vlan 3
SW1(config-vlan)#name V3                        //创建 VLAN3，并重命名为 V3
SW1(config-vlan)# end
SW1#show vlan                                   //查看 VLAN 信息

VLAN Name                            Status    Ports
---- -------------------------------- --------- -------------------------------
1    default                          active    Fa0/1, Fa0/2, Fa0/3, Fa0/4
                                                Fa0/5, Fa0/6, Fa0/7, Fa0/8
                                                Fa0/9, Fa0/10, Fa0/11, Fa0/12
                                                Fa0/13, Fa0/14, Fa0/15, Fa0/16
                                                Fa0/17, Fa0/18, Fa0/19, Fa0/20
                                                Fa0/21, Fa0/22, Fa0/23, Fa0/24
2    V2                               active
3    V3                               active
1002 fddi-default                     act/unsup
1003 token-ring-default               act/unsup
1004 fddinet-default                  act/unsup
1005 trnet-default                    act/unsup
……
```

从上面输出可见，VLAN1 是交换机上默认存在的 VLAN，交换机的端口在默认情况下都属于 VLAN 1；另外，1002~1005 号 VLAN 是交换机固定存在的 VLAN，一般情况下不会使用。

已创建了 V2 和 V3 两个 VLAN，但还没有端口属于这两个 VLAN。

删除 VLAN 的方法是：在全局模式下，用 "no vlan VLAN-id" 删除 VLAN，VLAN-id 为 VLAN 的编号。

另外，还有一种创建 VLAN 的方式，是在 VLAN 数据库模式下创建，但现在较少使用这种方式：

```
SW1#vlan database            //进入 VLAN 数据库模式
% Warning: It is recommended to configure VLAN from config mode, as VLAN database mode is being
```

deprecated. Please consult user documentation for configuring VTP/VLAN in config mode.
//提示不推荐使用这种方式，建议在全局模式下创建 VLAN

//下面按图 20-1，将端口分配到 VLAN 中：
SW1#conf t
SW1(config)#int f0/1
SW1(config-if)#switchport mode access　　//将 f0/1 端口设置为 access 模式，由于端口的模式默
　　　　　　　　　　　　　　　　　　　　　//认为 access，因此此语句可不配置
SW1(config-if)#switchport access vlan 2　　//将 f0/1 加入到 VLAN2 中
SW1(config-if)#int f0/2
SW1(config-if)#switchport access vlan 3
SW1(config-if)#end
SW1#show vlan　　　　　　　　　　　　　　//再查看 VLAN 信息

VLAN	Name	Status	Ports
1	default	active	Fa0/3, Fa0/4, Fa0/5, Fa0/6
			Fa0/7, Fa0/8, Fa0/9, Fa0/10
			Fa0/11, Fa0/12, Fa0/13, Fa0/14
			Fa0/15, Fa0/16, Fa0/17, Fa0/18
			Fa0/19, Fa0/20, Fa0/21, Fa0/22
			Fa0/23, Fa0/24
2	V2	active	Fa0/1
3	V3	active	Fa0/2
1002	fddi-default	act/unsup	
1003	token-ring-default	act/unsup	
1004	fddinet-default	act/unsup	
1005	trnet-default	act/unsup	

……

从输出可见，已经将 Fa0/1 分配到 V2，Fa0/2 分配到 V3 中了。

（2）配置交换机 SW2

Switch>en
Switch#conf t
Switch(config)#host SW2
SW2(config)#vlan 2
SW2(config-vlan)#name V2
SW2(config-vlan)#vlan 3
SW2(config-vlan)#name V3
SW2(config-vlan)#exit
SW2(config)#int range f0/3-4　　　　　　//用 int range 命令可将多个端口一起配置 VLAN，
　　　　　　　　　　　　　　　　　　　//要求是一起配置的端口应划分到同一 VLAN 中
SW2(config-if-range)#switchport access vlan 2　　//将 f0/3 和 f0/4 都加入到 VLAN2 中
SW2(config-if-range)#exit
SW2(config)#int f0/5
SW2(config-if)#switchport access vlan 3
SW2(config-if)#end

五、实验总结

1．测试各 PC 间的连通性

用 PC1 ping PC2，PC1 ping PC3，PC1 ping PC5，PC3 ping PC4……可以发现，除了 PC3 可以 ping 通 PC4 之外，其余的均不能 ping 通。这里需要先了解如下两个概念。

（1）VLAN 间路由

使用 VLAN 可将一个物理网内的多台 PC 划分不同的逻辑网，属于不同 VLAN 的 PC 之间需要通信，需要配置能实现在这些 VLAN 间进行路由转发的设备。

在本实验中，如 PC1 与 PC2 之间，分别属于 VLAN2 和 VLAN3，由于没有配置 VLAN 间的路由转发，因此不能 ping 通。

（2）trunk 模式（干道模式、主干模式）

交换机的 trunk 模式端口，在不同的 VLAN 流量经过时，会将 VLAN 号插入到数据包的前面，即为数据包打上 VLAN 标记，以区分数据包属于哪个 VLAN。

trunk 模式配置用在交换机与交换机相连的端口上，或者使用单臂路由的时候，与路由器相连的交换机端口应设置为 trunk 模式。

在本实验中，如 PC1 与 PC3，虽然都划分到了 VLAN2，但由于两交换机相连的端口没设置为 trunk 模式，因此不能 ping 通。

2．配置 trunk 端口

在 SW1 上：

  SW1(config)#int f0/24
  SW1(config-if)#switchport mode trunk　　　　　　//将 f0/10 端口设置为 trunk 模式
  SW1(config-if)#switchport trunk encapsulation dot1q　　//将 f0/24 的封装协议设为 802.1q，现在的
  　　　　　　　　　　　　　　　　　　　　　　　　//交换机大多默认封装为 802.1q，所以可
  　　　　　　　　　　　　　　　　　　　　　　　　//以不配置本语句（只有在不确定设备状
  　　　　　　　　　　　　　　　　　　　　　　　　//况时进行配置）。

同样，在 SW2 上：

  SW1(config)#int f0/24
  SW1(config-if)#switchport mode trunk

然后，用 PC1 ping PC3 或 PC4，PC2 ping PC5，都可以 ping 通。

3．设置干道上允许传输的 VLAN

在默认情况下，在 VLAN 干道上允许传输所有 VLAN，但可通过配置，只允许通过指定的 VLAN：

  SW1(config-if)#switchport trunk allowed vlan ?
   WORD　　　VLAN IDs of the allowed VLANs when this port is in trunking mode
   　　　　　　　　　　　　　　　　　　　　　　　//指定允许通过的 VLAN
   add　　　　add VLANs to the current list　　　　　//添加允许通过的 VLAN
   all　　　　　all VLANs　　　　　　　　　　　　//允许所有的 VLAN
   except　　　all VLANs except the following　　　　//不允许传输的 VLAN
   none　　　 no VLANs　　　　　　　　　　　　//不允许传输所有 VLAN

remove　remove VLANs from the current list　　//从已允许的 VLAN 列表中去除

在 SW1 上添加配置：

SW1(config-if)#switchport trunk allowed vlan 1,2

然后，再从 PC2 ping PC5，即测试 VLAN3 的两主机连通性，发现已不能 ping 通；而 VLAN2 中的 PC1 ping PC3 或 PC4 仍能 ping 通。

在 SW1 上添加配置，将 VLAN3 添加到允许的 VLAN 列表：

SW1(config-if)#switchport trunk allowed vlan add 3

此时再测试 VLAN3 的 PC2~PC5 的连通性，已能 ping 通了。

前面提到过：VLAN2 中的主机与 VLAN3 中的主机仍不能 ping 通，原因是不同的 VLAN 间通信，需要由三层设备来提供路由，下一个实验将讲述实现 VLAN 间路由的几种方式。

实验 21　VLAN 间路由

一、实验要求
- 掌握基于三层交换机的 VLAN 间路由；
- 掌握基于路由器物理接口的 VLAN 间路由；
- 掌握基于路由器子接口的 VLAN 间路由。

二、实验说明
　　实现不同的 VLAN 间主机的互访，从原理上讲，也需要像物理局域网那样，需要通过路由转发才能实现。但是，从具体的实现上讲，实现 VLAN 主机的互访，又与物理局域网互访有一定区别，实现 VLAN 间主机互访的方式有三种：一是通过三层交换机的路由功能来实现，二是通过路由器的物理接口来实现，三是通过路由器的子接口来实现，本实验将讲述这三种 VLAN 间的路由功能。

三、基于三层交换机的 VLAN 间路由
　　实验拓扑如图 21-1 所示。

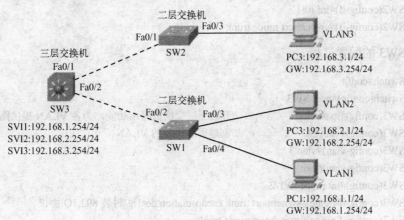

图 21-1　基于三层交换机的 VLAN 间路由实验拓扑

首先按图 21-1 所示，配置好各 PC 的 IP 地址及网关，以 PC1 为例，IP 配置如图 21-2 所示。

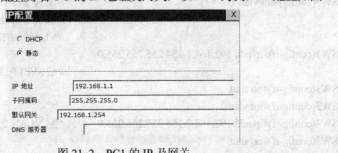

图 21-2　PC1 的 IP 及网关

然后配置各交换机。

（1）SW1 的配置过程

```
Switch#conf t
Switch(config)#host SW1
SW1(config)#vlan 2
SW1(config-vlan)#vlan 3
SW1(config-vlan)#exit
SW1(config)#int f0/4
SW1(config-if)# switchport access vlan 1
SW1(config-if)#int f0/3
SW1(config-if)# switchport access vlan 2
SW1(config-if)#int f0/2
SW1(config-if)#switchport mode trunk
```

（2）SW2 的配置过程

```
Switch#conf t
Switch(config)#host SW2
SW2(config)#vlan 3
SW2(config-vlan)#exit
SW2(config)#int f0/3
SW2(config-if)#switchport access vlan 3
SW2(config-if)#int f0/1
SW2(config-if)#switchport mode trunk
```

（3）SW3 的配置过程

```
Switch#conf t
Switch(config)#host SW3
SW3(config)#ip routing                              //开启三层交换机路由功能，为各 VLAN 提供路由转发
SW3(config)#vlan 2                                  //创建需要转发的 VLAN
SW3(config-vlan)#vlan 3
SW3(config-vlan)#exit
SW3(config)#int range f0/1-2
SW3(config-if-range)#switchport trunk encapsulation dot1q//封装 802.1Q 协议
SW3(config-if-range)#switchport mode trunk
SW3(config-if-range)#exit
SW3(config)#int vlan 1                              //配置 SVI 1，它是个虚拟接口，类
                                                    //似于路由器的一个端口，用于作为
                                                    //VLAN 1 的网关
SW3(config-if)#ip add 192.168.1.254 255.255.255.0
                                                    //给 SVI 1 配 IP 地址，即 VLAN 1 的网关地址
SW3(config-if)#no shut
SW3(config-if)#int vlan 2
SW3(config-if)#ip add 192.168.2.254 255.255.255.0   //配 VLAN 2 的网关地址
SW3(config-if)#no shut
SW3(config-if)#int vlan 3
```

```
SW3(config-if)#ip add 192.168.3.254 255.255.255.0        //配 VLAN 3 的网关地址
SW3(config-if)#no shut
```

实验总结：

配置完成后，在不同 VLAN 间的 PC 相互 ping 测试，都可以正常 ping 通。

在启用了三层交换机的路由功能后，这个三层交换机就相当于一个路由器，在这个三层交换机上为每个 VLAN 配置了虚拟接口及地址，这些虚拟接口相当于路由器的端口，当虚拟接口收到数据包后，三层交换机就进行路由选择，根据数据包的目的 IP 地址，选择正确的虚拟接口转发出去。

四、基于路由器物理接口的 VLAN 间路由

实验拓扑如图 21-3 所示。

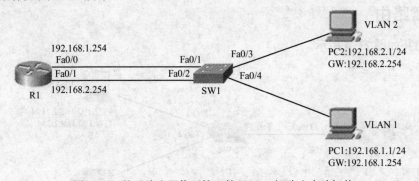

图 21-3 基于路由器物理接口的 VLAN 间路由实验拓扑

首先按照图 21-3 所示配置各 PC 的 IP 及网关，然后配置交换机 SW1 和路由器 R1。

（1）SW1 的配置过程

```
Switch#conf t
Switch(config)#host SW1
SW1(config)#vlan 2
SW1(config-vlan)#exit
SW1(config)#int f0/4
SW1(config-if)#switchport access vlan 1
SW1(config-if)#int f0/3
SW1(config-if)#switchport access vlan 2
SW1(config-if)#int f0/1
SW1(config-if)#switchport access vlan 1        //与路由器相连的端口 f0/2 划分给 VLAN 1
SW1(config-if)#int f0/2
SW1(config-if)#switchport access vlan 2        //与路由器相连的端口 f0/1 划分给 VLAN 2
```

（2）R1 的配置过程

```
Router#conf t
Router(config)#host R1
R1(config)#int f0/0
R1(config-if)#ip add 192.168.1.254 255.255.255.0    //在 f0/0 上配置 VLAN 1 的网关地址，用于转
                                                    //发 VLAN 1 流量
```

R1(config-if)#no shut
R1(config-if)#int f0/1
R1(config-if)#ip add 192.168.2.254 255.255.255.0   //在 f0/1 上配置 VLAN 2 的网关地址，用于转
                                                     //发 VLAN 2 流量
R1(config-if)#no shut

实验总结：

最后在 PC1 和 PC2 上进行连通性测试，可正常 ping 通。

通过路由器物理接口来实现 VLAN 间路由，这与路由器实现物理局域网间的路由相同，就是路由器的接口配置在不同的网络中，作为该网络的网关地址。

如果有多个 VLAN 间需要路由，则需要占用相同数量的物理端口，这将大大增加 VLAN 间路由的成本，因此可采用下一个实验的方法：单臂路由。

五、单臂路由

实验拓扑如图 21-4 所示。

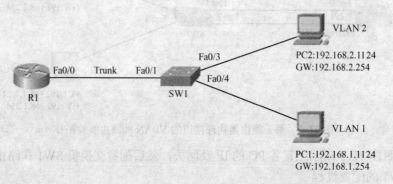

图 21-4 单臂路由实验拓扑

首先按照图 21-4 所示配置各 PC 的 IP 及网关，然后配置交换机 SW1 和路由器 R1。

（1）SW1 的配置过程

Switch#conf t
Switch(config)#host SW1
SW1(config)#vlan 2
SW1(config-vlan)#exit
SW1(config)#int f0/4
SW1(config-if)#switchport access vlan 1
SW1(config-if)#int f0/3
SW1(config-if)#switchport access vlan 2
SW1(config-if)#int f0/1
SW1(config-if)#switchport mode trunk   //将与路由器 R0 相连的端口配置成 trunk 模式

（2）R1 的配置过程

Router#conf t
Router(config)#host R1
R1(config)#int f0/0   //进入路由器的物理端口 f0/0，此端口是与交换机

```
R1(config-if)#no shut                              //需要在物理端口模式下打开
R1(config-if)#int f0/0.1                           //配置 f0/0 端口的第 1 个子接口
R1(config-subif)#encapsulation dot1q 1             //配置封装协议，并作为 VLAN 1 的网关
R1(config-subif)#ip add 192.168.1.254 255.255.255.0   //配置 VLAN 1 的网关 IP
R1(config-if)#int f0/0.2                           //配置 f0/0 端口的第 2 个子接口
R1(config-subif)#encapsulation dot1q 2
R1(config-subif)#ip add 192.168.2.254 255.255.255.0
```
（上方注释：//SW1 干道相连的端口）

实验总结：

① 在配置子接口时，应先配置封装协议，后配置 IP 地址，否则会报错。

② 需要为每个 VLAN 分别创建一个子接口，并配置相应的网关地址。

③ 子接口（Subinterface）是将一个物理接口（interface）虚拟出来的多个逻辑接口。相对子接口而言，所在的物理接口称为主接口。每个子接口从功能、作用上来说，与每个物理接口是没有任何区别的，它的出现打破了每个设备存在物理接口数量有限的局限性。一个物理接口可以配置多个子接口（0～4294967295 个），子接口性能受主接口物理性能限制，数量越多，各子接口性能越差。

④ 当存在多个 VLAN 的网络时，无法使用单台路由器的一个物理接口实现 VLAN 间通信，通过在一个物理接口上划分多个子接口的方式，从而实现多个 VLAN 间的路由和通信。

⑤ 与子接口相比，物理接口的性能更好，可将子接口配置在多个物理接口上，以减轻 VLAN 流量之间竞争带宽的现象。

⑥ 子接口的分类分为点到点子接口和点到多点子接口。

实验 22 VTP 的配置

一、实验要求

- 理解并掌握 VTP 修正号的作用及解决方法；
- 理解 VTP 配置中的相关概念；
- 掌握 VTP 的配置方法。

二、实验说明与实验拓扑

在 VTP 配置中，需要确保 VTP 域名一致、口令一致、版本一致、模式配置正确以及处理好修正号问题，在这几个方面中，修正号问题是最重要的，因为前几个方面对 VTP 的配置来说，最多也就是配置了 VTP 不能正常学习到 VLAN，但修正号问题解决不好，可能对整个网络带来崩溃性的灾难——以前整个网络配置的 VLAN 信息遭到破坏。

下面先就 VTP 修正号问题配置举例，如图 22-1 所示。

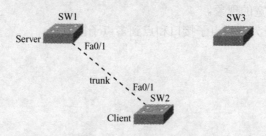

图 22-1 VTP 修正号 1

在图 22-1 中，SW1 和 SW2 表示网络中已有的交换机，SW3 是新加入的交换机，其中 SW1 是 Server 模式，SW2 是 Client 模式。

三、实验过程与实验总结

（1）SW2 的配置过程

```
Switch>en
Switch#conf t
Switch(config)#host SW2
SW2(config)#vtp mode client            //将 SW2 的 VTP 模式设为 Client 模式
Device mode already VTP CLIENT.        //提示模式已经设为 Client 模式了
SW2(config)#int f0/1
SW2(config-if)#switchport mode trunk   //将 f0/1 端口设为 trunk 模式
```

（2）SW1 的配置及调试过程

```
Switch>en
Switch#conf t
Switch(config)#host SW1
SW1(config)#vtp domain VTP1           //配置 VTP 域名为 VTP1
SW1(config)#vtp mode Server           //Server 模式是默认的，可以不配此句
SW1(config)#int f0/1
SW1(config-if)#switchport mode trunk  //各交换机相连的端口应设为 trunk 模式
SW1(config-if)#exit
SW1(config)#vlan 2                    //在 Server 模式交换机上配置 VLAN
SW1(config-vlan)#vlan 3
SW1(config-vlan)#vlan 4
SW1(config-vlan)#end
```

（3）查看 SW1 的 VLAN 信息

```
SW1#show vlan

VLAN Name                        Status      Ports
---- ---------------------------- ---------  -------------------------------
1    default                      active     Fa0/2, Fa0/3, Fa0/4, Fa0/5
                                             Fa0/6, Fa0/7, Fa0/8, Fa0/9
                                             Fa0/10, Fa0/11, Fa0/12, Fa0/13
                                             Fa0/14, Fa0/15, Fa0/16, Fa0/17
                                             Fa0/18, Fa0/19, Fa0/20, Fa0/21
                                             Fa0/22, Fa0/23, Fa0/24
2    VLAN0002                     active
3    VLAN0003                     active
4    VLAN0004                     active
1002 fddi-default                 act/unsup
1003 token-ring-default           act/unsup
1004 fddinet-default              act/unsup
1005 trnet-default                act/unsup
……      //可见在 SW1 上已新创建了三个 VLAN
```

（4）查看 SW1 的 VTP 状态

```
SW1#show vtp status
VTP Version                    : 2          //该 VTP 支持版本 2
Configuration Revision         : 3          //因为创建了三个 VLAN，对 VTP 修改了 3
                                            //次，因此修正号为 3
Maximum VLANs supported locally: 255        //当前交换机支持的最大 VLAN 数量
Number of existing VLANs       : 8          //当前已有 8 个 VLAN（其中有 5 个默认的）
VTP Operating Mode             : Server     //交换机模式是 Server 模式
VTP Domain Name                : VTP1       //VTP 域名为 VTP1
VTP Pruning Mode               : Disabled   //未启用 VTP 修剪功能
VTP V2 Mode                    : Disabled   //未启用版本 2，当前是版本 1
VTP Traps Generation           : Disabled
MD5 digest                     : 0x26 0x81 0x2E 0x65 0x00 0xEF 0x15 0x03
```

```
Configuration last modified by 0.0.0.0 at 3-1-93 00:01:21
Local updater ID is 0.0.0.0 (no valid interface found)
```

(5) 查看 SW2 的 VLAN 学习情况

```
SW2#show vlan

VLAN Name                        Status     Ports
---- ---------------------------- --------- -------------------------------
1    default                      active    Fa0/2, Fa0/3, Fa0/4, Fa0/5
                                            Fa0/6, Fa0/7, Fa0/8, Fa0/9
                                            Fa0/10, Fa0/11, Fa0/12, Fa0/13
                                            Fa0/14, Fa0/15, Fa0/16, Fa0/17
                                            Fa0/18, Fa0/19, Fa0/20, Fa0/21
                                            Fa0/22, Fa0/23, Fa0/24
2    VLAN0002                     active
3    VLAN0003                     active
4    VLAN0004                     active
1002 fddi-default                 act/unsup
1003 token-ring-default           act/unsup
1004 fddinet-default              act/unsup
1005 trnet-default                act/unsup
……
```

可见，SW2 已从 SW1 上学习到了 VLAN 信息。

(6) 查看 SW2 的 VTP 状态

```
SW2#show vtp status
VTP Version                     : 2
Configuration Revision          : 3             //修正号为 3，是从 SW1 上传过来的
Maximum VLANs supported locally : 255
Number of existing VLANs        : 8
VTP Operating Mode              : Client        //模式为 Client
VTP Domain Name                 : VTP1          //域名是从 Server 上通告的 VTP 中学到的
VTP Pruning Mode                : Disabled
VTP V2 Mode                     : Disabled
VTP Traps Generation            : Disabled
MD5 digest                      : 0x09 0xF4 0x41 0xCA 0xAF 0x6A 0xDE 0x91
Configuration last modified by 0.0.0.0 at 3-1-93 00:32:40
```

(7) 在 SW1 上，通过修改 VTP 域名使修正号为 0

```
SW1(config)#vtp domain VTP2
Changing VTP domain name from VTP1 to VTP2
SW1(config)#end
SW1#show vtp status
VTP Version                     : 2
```

```
Configuration Revision              : 0                //SW1 修改域名后,修正号变成了 0
Maximum VLANs supported locally     : 255
Number of existing VLANs            : 8
VTP Operating Mode                  : Server
VTP Domain Name                     : VTP2
VTP Pruning Mode                    : Disabled
VTP V2 Mode                         : Disabled
VTP Traps Generation                : Disabled
MD5 digest                          : 0x5B 0x5D 0xED 0x97 0xDC 0x63 0xB6 0xC5
Configuration last modified by 0.0.0.0 at 3-1-93 00:01:21
Local updater ID is 0.0.0.0 (no valid interface found)
SW1#00:03:27 %DTP-5-DOMAINMISMATCH: Unable to perform trunk negotiation on port Fa0/1
because of VTP domain mismatch.
SW1#00:03:57 %DTP-5-DOMAINMISMATCH: Unable to perform trunk negotiation on port Fa0/1
because of VTP domain mismatch.
       ……    //不停地提示:"不能进行干道协商,因为 VTP 域名不一致"。
```

把 VTP 域名改回 VTP1:

```
SW1(config)#vtp domain VTP1
Changing VTP domain name from VTP2 to VTP1      //提示域名改成了 VTP1
```

(8) 把 SW2 的模式改为透明模式,也可以使修正号变为 0

```
SW2(config)#vtp mode transparent                //改为透明模式
SW2#show vtp status
VTP Version                         : 2
Configuration Revision              : 0                //可见,修正号变成了 0
Maximum VLANs supported locally     : 255
Number of existing VLANs            : 8
VTP Operating Mode                  : Transparent
VTP Domain Name                     : VTP1
VTP Pruning Mode                    : Disabled
VTP V2 Mode                         : Disabled
VTP Traps Generation                : Disabled
MD5 digest                          : 0x90 0x4E 0xB2 0x29 0xEB 0xE4 0x24 0x53
Configuration last modified by 0.0.0.0 at 3-1-93 01:23:57
```

☞说明:

> 透明模式(transparent)的交换机可以创建、修改和删除 VLAN,但这些信息只对本交换机有效,不会通告其他交换机,也不会接收其他交换机的 VTP 通告来的 VLAN 信息。在这种模式下的交换机可以通过 Trunk 链路转发 VTP 信息。

(9) 在原有网络中加入交换机

如果 SW3 是一台新的交换机,其修正号为 0,将其加入网络后,设置为 Client 模式,整个网络的 VTP 工作将一切正常,它可以从 SW1 处通过 VTP 学到 VLAN 的相关信息。

但是,如果 SW3 是一台已使用过的交换机,其修正号可能是一个比较大的值(大于原网络中 SW1 的修正号值),这种情况下,如果不修改修正号,则 SW3 加入后,将自动修改原网络中 VLAN 的设置,会对原网络造成比较大的破坏性,下面以这种情况进行实验。

① 首先将 SW3 的修正值增大:

```
Switch>en
Switch#conf t
Switch(config)#host SW3
SW3(config)#int f0/2
SW3(config-if)#switchport mode trunk
SW3(config-if)#end
SW3#show vtp status
VTP Version                         : 2
Configuration Revision              : 0           //初始修正号为 0
Maximum VLANs supported locally     : 255
Number of existing VLANs            : 5
VTP Operating Mode                  : Server
VTP Domain Name                     :
VTP Pruning Mode                    : Disabled
VTP V2 Mode                         : Disabled
VTP Traps Generation                : Disabled
MD5 digest                          : 0x7D 0x5A 0xA6 0x0E 0x9A 0x72 0xA0 0x3A
Configuration last modified by 0.0.0.0 at 0-0-00 00:00:00
Local updater ID is 0.0.0.0 (no valid interface found)
SW3#conf t
SW3(config)#vlan 2
SW3(config-vlan)#no vlan 2
SW3(config)#vlan 2
SW3(config-vlan)#no vlan 2
SW3(config)#vlan 2
SW3(config-vlan)#no vlan 2
SW3(config)#vlan 2
SW3(config-vlan)#no vlan 2       //反复增删 VLAN,使修正号值增大,VLAN 没有增加
SW3(config)#end
SW3#show vtp status
VTP Version                         : 2
Configuration Revision              : 8           //此时修正号已变成了 8
Maximum VLANs supported locally     : 255
Number of existing VLANs            : 5           //仍然是默认的 5 个 VLAN
VTP Operating Mode                  : Server
VTP Domain Name                     :
VTP Pruning Mode                    : Disabled
VTP V2 Mode                         : Disabled
VTP Traps Generation                : Disabled
MD5 digest                          : 0x36 0xBD 0x5E 0x29 0xF4 0x66 0x7E 0x01
Configuration last modified by 0.0.0.0 at 3-1-93 01:40:13
Local updater ID is 0.0.0.0 (no valid interface found)
```

② 然后修改 SW3 的 VTP 模式和域名：

SW3(config)#vtp mode client
Setting device to VTP CLIENT mode.　　　　　//VTP 模式已改为 Client 模式
SW3(config)#vtp domain VTP1
Changing VTP domain name from NULL to VTP1　//VTP 域名已从空改为 VTP1

（10）把 SW3 与 SW2 相连

这相当于把一台修正号值大的交换机接入现有网络，如图 22-2 所示。

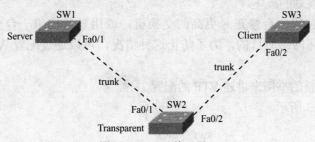

图 22-2　VTP 修正号 2

此时再查看 SW1 上的 VLAN 信息：

```
SW1#show vlan

VLAN Name                          Status     Ports
---- ------------------------------ --------- -------------------------------
1    default                        active    Fa0/2, Fa0/3, Fa0/4, Fa0/5
                                              Fa0/6, Fa0/7, Fa0/8, Fa0/9
                                              Fa0/10, Fa0/11, Fa0/12, Fa0/13
                                              Fa0/14, Fa0/15, Fa0/16, Fa0/17
                                              Fa0/18, Fa0/19, Fa0/20, Fa0/21
                                              Fa0/22, Fa0/23, Fa0/24
1002 fddi-default                   act/unsup
1003 token-ring-default             act/unsup
1004 fddinet-default                act/unsup
1005 trnet-default                  act/unsup
……
```

可见，原 SW1 上的 VLAN 已没有了，再查看 SW1 上的 VTP 信息：

```
SW1#show vtp status
VTP Version                     : 2
Configuration Revision          : 8         //修正号变成了 8，是从 SW3 上传过来的
Maximum VLANs supported locally : 255
Number of existing VLANs        : 5
VTP Operating Mode              : Server
VTP Domain Name                 : VTP1
VTP Pruning Mode                : Disabled
VTP V2 Mode                     : Disabled
VTP Traps Generation            : Disabled
```

127

MD5 digest   : 0x17 0x66 0x78 0xC2 0xF5 0x3C 0xFA 0xB9
Configuration last modified by 0.0.0.0 at 3-1-93 03:34:43
Local updater ID is 0.0.0.0 (no valid interface found)

通过这个实验证明了当原网络中加入的交换机如果具有较大的 VTP 修正值时，会对原网络的 VLAN 信息进行破坏，因此在新加入交换机到原网络中时，应特别注意修正值问题，解决的方法就是前面讲的，在接入网络之前：把 VTP 域名改成其他域名，再改回来，或者把 VTP 模式改为透明模式，再改回来，都可以使修正值清 0，然后再接入网络，避免了对原有网络 VLAN 规划的破坏。

如果网络黑客使用一台修正号更高的交换机，或用软件模拟一台交换机接入现有网络，就可以达到破坏网络的目的，为了防止这种情况，可以通过配置 VTP 密码来阻止黑客的破坏。

（11）结合较完整的举例来讲述 VTP 的配置

拓扑图如图 22-3 所示。

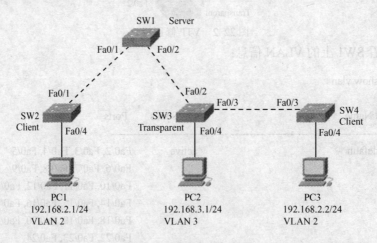

图 22-3 VTP 配置拓扑图

① 首先配置透明模式的交换机 SW3：

```
Switch>en
Switch#conf t
Switch(config)#host SW3
SW3(config)#int f0/2
SW3(config-if)#switchport mode tr
SW3(config-if)#int f0/3
SW3(config-if)#switchport mode tr
SW3(config-if)#exit
SW3(config)#vtp domain wgh
SW3(config)#vtp mode transparent
SW3(config)#vtp password ccna          //配置 VTP 密码，密码对大小写敏感
SW3(config)#vlan 3                     //在透明模式的交换机上配置 VLAN 3
```

② 配置 Server 模式交换机 SW1：

```
Switch>en
Switch#conf t
Switch(config)#host SW1
SW1(config)#int f0/1
SW1(config-if)#switchport mode tr
SW1(config-if)#int f0/2
SW1(config-if)#switchport mode tr
SW1(config-if)#exit
SW1(config)#vtp domain wgh
SW1(config)#vtp password ccna        //整个 VTP 域要求密码一致
SW1(config)#vlan 2
SW1(config-vlan)#vlan 3
```

③ 配置 Client 模式交换机 SW2 和 SW4：

```
Switch>en
Switch#conf t
Switch(config)#host SW2
SW2(config)#int f0/1
SW2(config-if)#switchport mode trunk
SW2(config-if)#exit
SW2(config)#vtp mode client
SW2(config)#vtp password ccna
SW2(config)#vtp domain wgh
Domain name already set to wgh.        //提示域名已经是 wgh 了
```

为什么没有配域名，而域名会自动配置呢？因为当交换机的域名为空时，如果它收到的 VTP 通告中带有域名信息，则会自动把收到的通告中的域名改为自己的域名；但是，如果 VTP 域名不为空，则不会自动更改域名了。

④ 然后查看 SW2 的 VLAN 信息，是否与 SW1 同步：

```
SW2#show vlan

VLAN Name                             Status    Ports
---- -------------------------------- --------- -------------------------------
1    default                          active    Fa0/2, Fa0/3, Fa0/5, Fa0/6
                                                Fa0/7, Fa0/8, Fa0/10, Fa0/11
                                                Fa0/12, Fa0/13, Fa0/14, Fa0/15
                                                Fa0/16, Fa0/17, Fa0/18, Fa0/19
                                                Fa0/20, Fa0/21, Fa0/22, Fa0/23
                                                Fa0/24
2    VLAN0002                         active    Fa0/4
3    VLAN0003                         active
1002 fddi-default                     act/unsup
1003 token-ring-default               act/unsup
1004 fddinet-default                  act/unsup
1005 trnet-default                    act/unsup
```

可见，SW2 已从 SW1 上同步了 VTP 信息。将端口指定到 VLAN：

    SW2(config)#int f0/4
    SW2(config-if)#switchport access vlan 2

⑤ 对 SW4 的配置：

    Switch>en
    Switch#conf t
    Switch(config)#host SW4
    SW4(config)#int f0/3
    SW4(config-if)#switchport mode trunk
    SW4(config-if)#exit
    SW4(config)#vtp password ccna
    SW4(config)#vtp domain wgh
    Domain name already set to wgh.   //与 SW2 一样，提示 VTP 域名已经是 wgh 了
    SW4(config)#vtp mode client
    SW4(config)#int f0/4
    SW4(config-if)#switchport access vlan 2

然后，按图 22-3 所示，配置好各 PC 的 IP 地址，并进行 ping 测试，发现 PC1 不能 ping 通 PC 3，原因是在 SW3 上，只添加了 VLAN 3，没有 VLAN 2，则 SW3 就不会转发 VLAN 2 的流量，因此，需要在 SW3 上添加 VLAN 2，这样就可以 ping 通了。

如果要使 SW3 上的 PC2 能与 PC1、PC3 通信，这需要用到前面讲的 VLAN 间路由才可以。

（12）VTP 裁剪功能

VTP Pruning（VTP 裁剪）是 VTP 的一个功能，VTP 通过裁剪来减少没有必要扩散的通信量，来提高中继链路的带宽利用率，VTP 裁剪默认是关闭状态。

默认情况下，某个 VLAN 的广播包会通过干道传输到所有交换机，即使该交换机上没有这个 VLAN 的端口也会从干道收到此广播包，如果配置了 VTP 裁剪后，则该交换机不会接收该 VLAN 的广播包，配置 VTP 裁剪的命令：

    Switch(config)#vtp pruning

实验 23 STP 的配置

一、实验要求
- 理解 STP 的工作原理；
- 掌握修改 STP 树的方法；
- 掌握 STP 负载均衡的配置；
- 掌握 RSTP 的配置。

二、实验说明
每个 VLAN 实例都可以有独立的生成树结构。在网络中实现不同的 VLAN 在多个交换机之间负载均衡是很有实际意义的。本实验讲述 STP 的工作原理和 RSTP 的配置，并着重讲了如何实现 STP 的负载均衡的配置过程。

三、实验拓扑
本实验使用拓扑结构如图 23-1 所示。

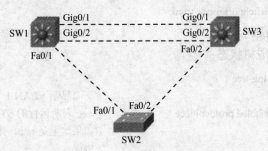

图 23-1 STP 配置实验拓扑

四、实验过程与实验总结
（1）交换机的基本配置

① SW1 的配置过程：

Switch>en
Switch#conf t
Switch(config)#host SW1
SW1(config)#vtp domain ccna   //SW1 作为 VTP Server 模式
SW1(config)#vlan 2
SW1(config-vlan)#exit
SW1(config)#int g0/1
SW1(config-if)#switchport trunk encapsulation dot1q//交换机间的链路配置为干道，传输 VLAN 数据
SW1(config-if)#switchport mode trunk
SW1(config-if)#int g0/2
SW1(config-if)#switchport trunk encapsulation dot1q
SW1(config-if)#switchport mode trunk

② SW2 的配置过程：

 Switch>en
 Switch#conf t
 Switch(config)#host SW2
 SW2(config)#vtp mode client   //SW2、SW3 作为 VTP Client 模式，可从 SW1
                  //学习 VLAN 及域名
 SW2(config)#int g0/1
 SW2(config-if)#switchport trunk encapsulation dot1q
 SW2(config-if)#switchport mode trunk
 SW2(config-if)#int g0/2
 SW2(config-if)#switchport trunk encapsulation dot1q
 SW2(config-if)#switchport mode trunk

③ SW3 的配置过程：

 Switch>en
 Switch#conf t
 Switch(config)#host SW3
 SW3(config)#vtp mode client
 SW3(config)#int f0/1
 SW3(config-if)#switchport mode trunk
 SW3(config-if)#int f0/2
 SW3(config-if)#switchport mode trunk

(2) 查看默认的生成树

① 查看 SW1 的生成树状态：

```
SW1#show spanning-tree
VLAN0001                                         //下面显示 VLAN 1 生成树状态
  Spanning tree enabled protocol ieee            //ieee 表明运行的 STP 协议是 IEEE 的 802.1d,
                                                 //如果这里是 rstp，则表明运行的是快速生成树
                                                 //协议
  Root ID    Priority    32769                   //根桥的优先级 32769，默认值 32768+VLAN 号
             Address     000C.CF99.EE26          //根桥的 MAC 地址
             Cost        4                       //从本交换机到根桥的 Cost 值
             Port        25(GigabitEthernet0/1)  //本交换机的第 25 端口（前有 24 个 100 兆端口）
             Hello Time 2 sec   Max Age 20 sec   Forward Delay 15 sec //各计时器时间

  Bridge ID  Priority    32769   (priority 32768 sys-id-ext 1)  //本机优先级
             Address     00D0.BC0B.39C0                         //本机 MAC 地址（比根桥 MAC 大）
             Hello Time  2 sec   Max Age 20 sec   Forward Delay 15 sec
             Aging Time  20

Interface        Role Sts Cost       Prio.Nbr Type
---------------- ---- --- ---------- -------- --------------------------------
Gi0/1            Root FWD 4          128.25   P2P
Gi0/2            Altn BLK 4          128.26   P2P       //本机上阻塞的端口是 G0/2 端口
Fa0/1            Desg FWD 19         128.1    P2P
```

在 Role（角色）下的 Root 表示根端口，Altn 表示轮换端口，Desg 是指定端口；Sts（状态）下的 FWD 表示转发状态，BLK 表示阻塞状态；Cost 下的值是该接口的开销值；Prio.Nbr 是接口的优先级；Type 下的 P2P 表示该接口是点对点类型。

```
VLAN0002                                    //下面显示 VLAN 2 生成树状态
  Spanning tree enabled protocol ieee
  Root ID    Priority    32770
             Address     000C.CF99.EE26
             Cost        4
             Port        25(GigabitEthernet0/1)
             Hello Time  2 sec   Max Age 20 sec   Forward Delay 15 sec

  Bridge ID  Priority    32770   (priority 32768 sys-id-ext 2)
             Address     00D0.BC0B.39C0
             Hello Time  2 sec   Max Age 20 sec   Forward Delay 15 sec
             Aging Time  20

  Interface       Role Sts Cost      Prio.Nbr Type
  ---------       ---- --- ----      -------- ----
  Gi0/1           Root FWD 4         128.25   P2P
  Gi0/2           Altn BLK 4         128.26   P2P    //与 VLAN 1 相同，阻塞 G0/2 端口
  Fa0/1           Desg FWD 19        128.1    P2P
```

默认情况下，交换机会对每个 VLAN 都生成一个单独的 STP 树，称为 PVST（Per VLAN Spanning Tree）。在 VLAN 2 的生成树状态信息中，除了优先级是 32770（32768+2）外，其余与 VLAN 1 相同。在没有对生成树做其他的配置时，生成树协议将对所有的 VLAN 进行相同的算法，其结果就是具有相同的根桥和相同的阻塞端口。

② 查看 SW2 的生成树状态：

```
SW2#show spanning-tree
VLAN0001
  Spanning tree enabled protocol ieee
  Root ID    Priority    32769
             Address     000C.CF99.EE26         //根桥的 MAC 地址
             This bridge is the root            //指明这个交换机为根桥
             Hello Time  2 sec   Max Age 20 sec   Forward Delay 15 sec

  Bridge ID  Priority    32769   (priority 32768 sys-id-ext 1)
             Address     000C.CF99.EE26         //这是本交换机的 MAC，就是根桥的
                                                //MAC 地址
             Hello Time  2 sec   Max Age 20 sec   Forward Delay 15 sec
             Aging Time  20

  Interface       Role Sts Cost      Prio.Nbr Type
  ---------       ---- --- ----      -------- ----
  Fa0/2           Desg FWD 19        128.2    P2P
  Gi0/1           Desg FWD 4         128.25   P2P
```

```
            Gi0/2              Desg FWD 4           128.26      P2P

    VLAN0002
    .......                //省略这部分输出，与 VLAN 1 输出完全一样
```

③ 查看 SW3 的生成树状态：

```
SW3#show spanning-tree
VLAN0001
Spanning tree enabled protocol ieee
    Root ID    Priority    32769
               Address     000C.CF99.EE26              //根桥的 MAC 地址
               Cost        19
               Port        2(FastEthernet0/2)
               Hello Time  2 sec   Max Age 20 sec   Forward Delay 15 sec

    Bridge ID  Priority    32769   (priority 32768 sys-id-ext 1)
               Address     0010.1169.96E9              //本交换机的 MAC 地址（比根桥
                                                       //MAC 大）
               Hello Time  2 sec   Max Age 20 sec   Forward Delay 15 sec
               Aging Time  20
    Interface          Role Sts Cost      Prio.Nbr Type
    ---------------- ---- --- --------- -------- --------------------------------
    Fa0/1              Altn BLK 19        128.1    P2P       //阻塞了
    Fa0/2              Root FWD 19        128.2    P2P


VLAN0002
.......                //省略这部分输出，与 VLAN 1 输出完全一样
    Interface          Role Sts Cost      Prio.Nbr Type
    ---------------- ---- --- --------- -------- --------------------------------
    Fa0/1              Altn BLK 19        128.1    P2P
    Fa0/2              Root FWD 19        128.2    P2P
```

交换机根桥选举的原则：BID 最小的作为根桥，BID=交换机优先级+ MAC 地址。

通过输出拓扑图中的三台交换机的 STP 状态可见，在优先级相同的情况下，SW2 的 MAC 值最小，SW2 被选为根桥。下面明确两个概念。

- 根端口：到达根桥的路径开销最小的那个端口；
- 指定端口：够接收其他交换机转发来的网络流量的那个端口。根桥上的所有端口都是指定端口，处于 Forward 状态。

确定根端口的规则：

- 一个交换机有两条或以上能到达根桥的路径，则选择累加成本最小的路径对应的那个端口为根端口；
- 在到达根桥的邻居交换机中，选择和 BID 最小的邻居交换机相连的端口为根端口；
- 如果多条路径都需经过同一邻居交换机，则选择拥有优先级值最小的端口为根端口；
- 选择端口物理编号最低的端口作为根端口。

根据上述规则，来解释 STP 产生的阻塞端口的原因，各链路的 Cost 值如图 23-2 所示。

根据第一条规则，SW3 到根桥 SW2 有两条路径，形成了环路，通过 f0/2，Cost 值是 19，通过 F0/1，Cost 值是 19+4=23，因此，根端口选 F0/2 为根端口，阻塞 F0/1 端口。

根据第四条规则，SW1 上的 G0/1 和 G0/2 到达根桥的 Cost 值都是 4，并形成了环路，此时选物理编号更低的 G0/1 端口为根端口，则阻塞 G0/2 端口。

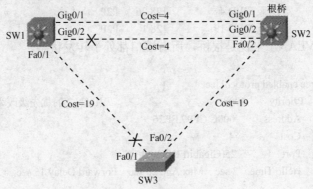

图 23-2 链路 Cost 值

在图 23-1 所示拓扑图中，SW1 和 SW2 是两台性能较好的三层交换机，可通过修改优先级，使它们在 VLAN 1 和 VLAN 2 间实现负载平衡。

把 SW1 配置为 VLAN 1 的根桥，指定 SW2 为 VLAN 1 的辅助根桥；同时，把 SW2 配置为 VLAN 2 的根桥，指定 SW1 为 VLAN 2 的辅助根桥。而 SW3 仍采用默认的优先级，其优先级值对每个 VLAN 来说都最大，则 SW3 永远不会成为根桥。

```
SW1(config)#spanning-tree vlan 1 priority 4096
          //配置交换机 SW1 对 VLAN 1 的优先级为 4096，在 SW2 和 SW3 上保持 VLAN
          //1 的优先级仍为 32768，这就使得 SW1 成了 VLAN 1 的根桥
SW1(config)#spanning-tree vlan 2 root secondary
          //配置交换机 SW1 为 VLAN 2 的辅助根桥
SW2(config)#spanning-tree vlan 2 priority 4096
          //配置交换机 SW2 对 VLAN 2 的优先级为 4096，在 SW1 和 SW3 上保持 VLAN
          //2 的优先级仍为 32768，这就使得 SW2 成了 VLAN 2 的根桥
SW2(config)#spanning-tree vlan 1 root secondary
          //配置交换机 SW1 为 VLAN 2 的辅助根桥
```

④ 查看各交换机的生成树状态：

● SW1 的生成树状态：

```
SW1#show spanning-tree
VLAN0001
   Spanning tree enabled protocol ieee
   Root ID    Priority    4097                  //VLAN 1 的优先级改为了 4097（4096+1）
              Address     00D0.BC0B.39C0
              This bridge is the root           //SW1 就是 VLAN 1 的根桥
              Hello Time  2 sec  Max Age 20 sec  Forward Delay 15 sec
   Bridge ID  Priority    4097    (priority 4096 sys-id-ext 1)   //本机就是 VLAN 1 的根桥，因此优
                                                                //先级相同
              Address     00D0.BC0B.39C0
              Hello Time  2 sec  Max Age 20 sec  Forward Delay 15 sec
```

```
                       Aging Time    20

     Interface         Role Sts Cost         Prio.Nbr Type
     ----------------  ---- --- --------  -------- --------------------
     Gi0/1             Desg FWD 4            128.25   P2P
     Gi0/2             Desg FWD 4            128.26   P2P
     Fa0/1             Desg FWD 19           128.1    P2P
```

可见，SW1 对 VLAN 1 来说是根桥，所有端口都处于转发状态。

```
     VLAN0002
       Spanning tree enabled protocol ieee
       Root ID     Priority     4098                    //VLAN 2 的优先级改为了 4098（4096+2）
                   Address      000C.CF99.EE26
                   Cost         4
                   Port         25(GigabitEthernet0/1)
                   Hello Time   2 sec   Max Age 20 sec    Forward Delay 15 sec

       Bridge ID   Priority    28674   (priority 28672 sys-id-ext 2)  //本交换机对 VLAN 2 改为了辅助
                                                                      //根桥，优先级改成了 28674
                   Address      00D0.BC0B.39C0
                   Hello Time   2 sec   Max Age 20 sec    Forward Delay 15 sec
                   Aging Time   20
     Interface         Role Sts Cost         Prio.Nbr Type
     ----------------  ---- --- --------  -------- --------------------
     Gi0/1             Root FWD 4            128.25   P2P
     Gi0/2             Altn BLK 4            128.26   P2P    //对 VLAN 2 来说，阻塞的是 G0/2 端口
     Fa0/1             Desg FWD 19           128.1    P2P
```

- SW2 的生成树状态：

```
     SW2#show spanning-tree
     VLAN0001
       Spanning tree enabled protocol ieee
       Root ID     Priority     4097                    //VLAN 1 的优先级 4097
                   Address      00D0.BC0B.39C0
                   Cost         4
                   Port         25(GigabitEthernet0/1)
                   Hello Time   2 sec   Max Age 20 sec    Forward Delay 15 sec

       Bridge ID   Priority    28673   (priority 28672 sys-id-ext 1)  //本交换机对 VLAN 1 改为了
                                                                      //辅助根桥，优先级改成了 28673
                   Address      000C.CF99.EE26
                   Hello Time   2 sec   Max Age 20 sec    Forward Delay 15 sec
                   Aging Time   20

     Interface         Role Sts Cost         Prio.Nbr Type
     ----------------  ---- --- --------  -------- --------------------
     Fa0/2             Desg FWD 19           128.2    P2P
```

```
    Gi0/1           Root FWD 4           128.25    P2P
    Gi0/2           Altn BLK 4           128.26    P2P      //对 VLAN 1 来说，阻塞的是 G0/2 端口


VLAN0002
  Spanning tree enabled protocol ieee
  Root ID    Priority    4098                              //VLAN 2 的优先级 4098
             Address     000C.CF99.EE26
             This bridge is the root
             Hello Time   2 sec  Max Age 20 sec   Forward Delay 15 sec

  Bridge ID  Priority    4098   (priority 4096 sys-id-ext 2)   //本机就是 VLAN 2 的根桥，因此优
                                                                //先级相同
             Address     000C.CF99.EE26
             Hello Time   2 sec  Max Age 20 sec   Forward Delay 15 sec
             Aging Time   20

Interface       Role Sts Cost      Prio.Nbr Type
--------------- ---- --- ---------  -------- --------------------------------
Fa0/2           Desg FWD 19          128.2    P2P
Gi0/1           Desg FWD 4           128.25   P2P
Gi0/2           Desg FWD 4           128.26   P2P
```

可见，SW2 对 VLAN 2 来说是根桥，所有端口都处于转发状态。

- SW3 的生成树状态：

```
SW3#show spanning-tree
VLAN0001
  Spanning tree enabled protocol ieee
  Root ID    Priority    4097
             Address     00D0.BC0B.39C0
             Cost        19
             Port        1(FastEthernet0/1)
             Hello Time   2 sec  Max Age 20 sec   Forward Delay 15 sec

  Bridge ID  Priority    32769  (priority 32768 sys-id-ext 1)
             Address     0010.1169.96E9
             Hello Time   2 sec  Max Age 20 sec   Forward Delay 15 sec
             Aging Time   20

Interface       Role Sts Cost      Prio.Nbr Type
--------------- ---- --- ---------  -------- --------------------------------
Fa0/1           Root FWD 19          128.1    P2P
Fa0/2           Altn BLK 19          128.2    P2P     //对 VLAN 1，阻塞了 F0/2 端口


VLAN0002
  Spanning tree enabled protocol ieee
  Root ID    Priority    4098
             Address     000C.CF99.EE26
```

```
              Cost          19
              Port          2(FastEthernet0/2)
              Hello Time    2 sec    Max Age 20 sec    Forward Delay 15 sec

     Bridge ID  Priority    32770    (priority 32768 sys-id-ext 2)
                Address     0010.1169.96E9
                Hello Time  2 sec    Max Age 20 sec    Forward Delay 15 sec
                Aging Time  20

     Interface        Role Sts Cost     Prio.Nbr Type
     ---------------- ---- --- -------- -------- --------------------------------
     Fa0/1            Altn BLK 19       128.1    P2P       //对 VLAN 2，阻塞了 F0/1 端口
     Fa0/2            Root FWD 19       128.2    P2P
```

分析上面的输出，可以发现，在 SW1 上对 VLAN 2 所形成的环路，阻塞的端口是 G0/2，在 SW2 上对 VLAN 1 所形成的环路，阻塞的端口也是 G0/2，其结果是 VLAN 1、VLAN 2 的数据需要在 SW1 和 SW2 上传输时，都是通过 G0/1 来实现的，这是不合理的。

可通过在端口上修改 VLAN 的端口优先级，来实现负载均衡：

```
SW1(config)#int g0/2                                      //对 G0/2 端口修改优先级
SW1(config-if)#spanning-tree vlan 1 port-priority 112
              //将 G0/2 端口对 VLAN 1 的优先级设为 112，其默认值是 128，值越小，优先级
              //越高，并要求按 16 的整数倍来设置。设置后，VLAN1 的流量将从 G0/2 上传输
```

另外，还可以通过修改 Cost 值来实现负载均衡：

```
SW1(config)#int g0/2
SW1(config-if)#spanning-tree vlan 2 cost 3
              //G0/2 的 Cost 值默认是 4，现改为 3，这将使得原来 VLAN 2 在 SW1 上的 G0/2
              //被阻塞，现在通过降低其 cost 值后，SW1 上的 G0/2 成为 VLAN 2 的根端口。
              //阻塞的端口变成了 SW1 上的 G0/1 端口。
```

注：通过修改 Cost 值来实现负载均衡的命令在"Cisco Packet Tracer"不支持。

⑤ 在各交换机上配置快速生成树协议（RSTP）。

由于标准 STP 的收敛时间大约是 30~50s，对于那些要求快速收敛的网络来说，这个速度就太慢了，快速生成树协议（Rapid Spanning Tree Protocol，RSTP）是由 IEEE 在参照 Cisco 制定的新标准基础上而制定出来的协议。

配置了 RSTP 后，一旦原有网络拓扑发生变化，可以在很短的时间内完成收敛过程，不会像 STP 那样需要花费 30~50s。

配置方法，就是在每个交换机上增加一条语句：

```
SW1(config)#spanning-tree mode rapid-pvst
SW2(config)#spanning-tree mode rapid-pvst
SW3(config)#spanning-tree mode rapid-pvst
```

配置完成后，可查看 RSTP 状态，以 SW1 为例：

```
SW1#show spanning-tree
VLAN0001
   Spanning tree enabled protocol rstp        //协议已不是 ieee，而是 RSTP
   ……    //后面与 STP 类似，省略
```

实验 24 二层交换机链路聚合

一、实验要求

（1）理解端口聚合的作用和特点；
（2）掌握端口聚合的配置方法。

二、实验说明

EtherChannel(以太通道)是由 Cisco 研发的，应用于交换机之间的多链路捆绑技术。它是两台交换机之间在物理上将多个端口连接起来，将多条链路聚合成一条逻辑链路，形成一个拥有较大宽带的链路，从而形成一条干路，增大链路带宽，可以实现均衡负载，并提供冗余链路，增强了网络的稳定性和安全性。在 EtherChannel 中，负载在各个链路上的分布可以根据源 IP 地址、目的 IP 地址、源 MAC 地址、目的 MAC 地址、源 IP 地址和目的 IP 地址组合，以及源 MAC 地址和目的 MAC 地址组合等来进行分布。两台交换机之间是否形成 EtherChannel 也可以用协议自动协商。目前有两个协商协议：PAgP 和 LACP，PAgP（Port Aggregation Protocol，端口汇聚协议）是 Cisco 私有的协议，而 LACP（Link Aggregation Control Protocol，链路汇聚控制协议）是基于 IEEE 802.3ad 的国际标准。

注意：先将两台交换机都配置完端口聚合后，再将两台交换机连接，如果先连线，后配置，会造成广播风暴。

三、实验拓扑

本实验的拓扑图如图 24-1 所示。

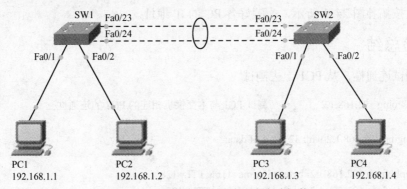

图 24-1 二层交换机的链路聚合实验拓扑

四、配置过程

（1）SW1 的配置过程

```
Switch>en
Switch#conf t
Switch(config)#host SW1
SW1(config)#vlan 10
SW1(config-vlan)#exit
SW1(config)#int range f0/1-10
SW1(config-if-range)#switchport access vlan 10      //将端口 1~10 指定给 VLAN10
SW1(config-if-range)#exit
SW1(config)#interface port-channel 1                //创建聚合链路 1
SW1(config-if)#switchport mode trunk                //定义聚合链路封装模式为 trunk 模式
SW1(config-if)#no shut
SW1(config-if)#exit
SW1(config)#int range f0/23-24
SW1(config-if-range)#end
```

(2) SW2 的配置过程

```
Switch>en
Switch#conf t
Switch(config)#host SW2
SW2(config)#vlan 10
SW2(config-vlan)#exit
SW2(config)#int range f0/1-10
SW2(config-if-range)#switchport access vlan 10
SW2(config-if-range)#exit
SW2(config)#interface port-channel 1
SW2(config-if)#switchport mode trunk
SW2(config-if)#no shut
SW2(config-if)#exit
SW2(config)#interface range f0/23-24
SW2(config-if-range)#channel-group 1 mode on
SW2(config-if-range)#end
```

然后，按拓扑图 24-1 所示，配置好各 PC 的 IP 地址。

五、实验总结

(1) 测试连通性，从 PC1 上去测试

```
PC>ping 192.168.1.2        //测试 PC1 与本交换机相连的 PC2 的连通性

Pinging 192.168.1.2 with 32 bytes of data:

Reply from 192.168.1.2: bytes=32 time=11ms TTL=128
Reply from 192.168.1.2: bytes=32 time=9ms TTL=128
Reply from 192.168.1.2: bytes=32 time=6ms TTL=128
Reply from 192.168.1.2: bytes=32 time=8ms TTL=128

Ping statistics for 192.168.1.2:
```

```
    Packets: Sent = 4, Received = 4, Lost = 0 (0% loss),
    Approximate round trip times in milli-seconds:
        Minimum = 6ms, Maximum = 11ms, Average = 8ms
PC>ping 192.168.1.3          //测试 PC1 与交换机 SW2 相连的 PC3 的连通性

Pinging 192.168.1.3 with 32 bytes of data:

Reply from 192.168.1.3: bytes=32 time=17ms TTL=128
Reply from 192.168.1.3: bytes=32 time=12ms TTL=128
Reply from 192.168.1.3: bytes=32 time=8ms TTL=128
Reply from 192.168.1.3: bytes=32 time=10ms TTL=128

Ping statistics for 192.168.1.3:
    Packets: Sent = 4, Received = 4, Lost = 0 (0% loss),
    Approximate round trip times in milli-seconds:
        Minimum = 8ms, Maximum = 17ms, Average = 11ms
```

可见，能正常通信。

（2）查看聚合链路信息

```
Switch#show etherchannel summary
Flags:   D - down         P - in port-channel
         I - stand-alone  s - suspended
         H - Hot-standby (LACP only)
         R - Layer3       S - Layer2
         U - in use       f - failed to allocate aggregator
         u - unsuitable for bundling
         w - waiting to be aggregated
         d - default port
Number of channel-groups in use: 1     //有 1 个以太网通道在使用
Number of aggregators:           1     //有 1 个聚合链路

Group  Port-channel  Protocol    Ports
------+-------------+-----------+-----------------------------------------

1      Po1(SU)       PAgP       Fa0/23(P) Fa0/24(P)
```

其中："SU"中的"S"代表的是二层以太网通道；"U"代表 UP，即通道在正常使用；"P"代表此接口参与了以太网通道。

（3）查看 EtherChannel 的模式

```
SW1(config-if-range)#channel-group 1 mode ?
                //查看在两台交换机之间动态协商建立以太网信道的方式。Cisco 设备支持端口
                //汇聚协议 PAgP 和链路聚合控制协议 LACP。
  active      Enable LACP unconditionally
                //active 是主动请求
  auto        Enable PAgP only if a PAgP device is detected
                //auto 是被动监听，等待被请求
```

```
desirable   Enable PAgP unconditionally
            //desirable 是用 Cisco 的 PAGP 协议，并且是主动协商
on          Enable Etherchannel only
            //on 是不用任何协议，强制开启以太网通道
passive     Enable LACP only if a LACP device is detected
            //passive 是被动监听，等待被请求
```

能形成 EtherChannel 的模式总结：

强制启用	on
on	✓

PAgP 协议	desirable	auto
desirable	✓	✓
auto	✓	✗

LACP 协议	active	passive
active	✓	✓
passive	✓	✗

实验 25 三层交换机链路聚合

一、实验要求

- 理解交换机三层链路聚合的作用和特点；
- 掌握链路聚合的配置方法。

二、实验说明

在前一个实验中，讲了二层链路聚合，但二层链路聚合不能提供 VLAN 间的互访，在本实验中，讲述交换机三层链路聚合，实现 VLAN 间的互访。

三、实验拓扑

本实验拓扑如图 25-1 所示。

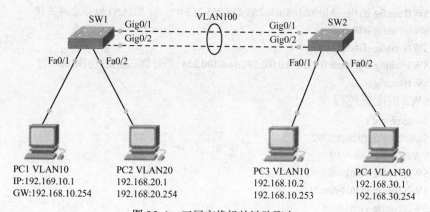

图 25-1 三层交换机的链路聚合

四、配置过程

（1）SW1 的配置过程
```
Switch#conf t
Switch(config)#host SW1
SW1(config)#vlan 10
SW1(config-vlan)#vlan 20
SW1(config-vlan)#vlan 100
```

```
SW1(config-vlan)#exit
SW1(config)#int f0/1
SW1(config-if)#switchport access vlan 10        //将 f0/1 端口指定给 VLAN10
SW1(config-if)#exit
SW1(config)#int f0/2
SW1(config-if)#switchport access vlan 20        //将 f0/2 端口指定给 VLAN10
SW1(config-if)#exit
SW1(config)#int range g0/1-2
SW1(config-if-range)#switchport access vlan 100 //将 g0/1，g0/2 端口指定给 VLAN100
SW1(config-if-range)#channel-group 1 mode desirable
                                                //g0/1，g0/2 端口进行链路聚合，配置为 PAgP 的
                                                //desirable 模式
SW1(config-if-range)#exit
SW1(config)#int port-channel 1
SW1(config-if)#switchport trunk encapsulation dot1q
SW1(config-if)#switchport mode trunk            //将聚合端口设置为 trunk 模式
SW1(config-if)#exit
SW1(config)#int vlan 10
SW1(config-if)#ip add 192.168.10.254 255.255.255.0   //给 VLAN10 指定网关 IP
SW1(config-if)#no shut
SW1(config-if)#int vlan 20
SW1(config-if)#ip add 192.168.20.254 255.255.255.0   //给 VLAN20 指定网关 IP
SW1(config-if)#no shut
SW1(config-if)#int vlan 100
SW1(config-if)#ip add 192.168.100.253 255.255.255.0  //给 VLAN100 指定网关 IP
SW1(config-if)#no shut
SW1(config-if)#exit
SW1(config)#ip route 0.0.0.0 0.0.0.0 192.168.100.254  //开启三层交换机路由功能
SW1(config)#
```

(2) SW2 的配置过程

```
Switch#conf t
Switch(config)#host SW2
SW2(config)#vlan 10
SW2(config-vlan)#vlan 30
SW2(config-vlan)#vlan 100
SW2(config-vlan)#exit
SW2(config)#int f0/1
SW2(config-if)#switchport access vlan 10
SW2(config-if)#int f0/3
SW2(config-if)#switchport access vlan 30
SW2(config)#int range g0/1-2
SW2(config-if-range)#switchport access vlan 100
SW2(config-if-range)#channel-group 1 mode desirable
SW2(config-if-range)#exit
SW2(config)#int port-channel 1
SW2(config-if)#switchport trunk encapsulation dot1q
```

```
SW2(config-if)#switchport mode trunk
SW2(config-if)#exit
SW2(config)#int vlan 10
SW2(config-if)#ip add 192.168.10.253 255.255.255.0
SW2(config-if)#no shut
SW2(config-if)#int vlan 30
SW2(config-if)#ip add 192.168.30.254 255.255.255.0
SW2(config-if)#no shut
SW2(config-if)#int vlan 100
SW2(config-if)#ip add 192.168.100.254 255.255.255.0
SW2(config-if)#no shut
SW2(config-if)#exit
SW2(config)#ip route 0.0.0.0 0.0.0.0 192.168.100.253
SW2(config)#
```

然后，按拓扑图 25-1 所示，配置好各 PC 的 IP 地址。

五、实验总结

1. 测试连通性

（1）从 PC1 ping SW2 的 PC3

```
PC>ping 192.168.10.2

Pinging 192.168.10.2 with 32 bytes of data:

Reply from 192.168.10.2: bytes=32 time=14ms TTL=128
Reply from 192.168.10.2: bytes=32 time=13ms TTL=128
Reply from 192.168.10.2: bytes=32 time=11ms TTL=128
Reply from 192.168.10.2: bytes=32 time=9ms TTL=128

Ping statistics for 192.168.10.2:
    Packets: Sent = 4, Received = 4, Lost = 0 (0% loss),
Approximate round trip times in milli-seconds:
    Minimum = 9ms, Maximum = 14ms, Average = 11ms
```

（2）从 PC2 ping SW2 的 PC4

```
PC>ping 192.168.30.1

Pinging 192.168.30.1 with 32 bytes of data:

Reply from 192.168.30.1: bytes=32 time=8ms TTL=126
Reply from 192.168.30.1: bytes=32 time=12ms TTL=126
Reply from 192.168.30.1: bytes=32 time=15ms TTL=126
Reply from 192.168.30.1: bytes=32 time=11ms TTL=126
Reply from 192.168.30.1: bytes=32 time=12ms TTL=126
Reply from 192.168.30.1: bytes=32 time=15ms TTL=126
```

Ping statistics for 192.168.30.1:
    Packets: Sent = 4, Received = 6, Lost = 4294967294 (4294967246% loss),
Approximate round trip times in milli-seconds:
    Minimum = 8ms, Maximum = 15ms, Average = 12ms

可见，在同一个 LAN 间的 PC 或不同 VLAN 间的 PC，都能相互通信。

2．关于 EtherChannel 的说明

① Cisco 最多允许 EtherChannel 绑定8个端口：如果是快速以太网，总带宽可达 1600Mbit/s；如果是 Gbit 以太网，总带宽可达 16Gbit/s。

② EtherChannel 不支持 10M 端口。

③ EtherChannel 编号只在本地有效，链路两端的编号可以不一样。

④ EtherChannel 默认使用 PAgP 协议。

⑤ EtherChannel 默认情况下是基于源 MAC 地址的负载平衡。

⑥ 一个 EtherChannel 内所有的端口都必须具有相同的端口速率和双工模式，LACP 只能是全双工模式。

⑦ channel-group 接口会自动继承最小物理接口，或最先配置的接口模式。

⑧ Cisco 的交换机不仅可以支持第二层 EtherChannel，还可以支持第三层 EtherChannel。

3．查看 EtherChannel 负载平衡方式

```
SW1#show etherchannel load-balance
EtherChannel Load-Balancing Configuration:
        src-mac

EtherChannel Load-Balancing Addresses Used Per-Protocol:
Non-IP: Source MAC address
  IPv4: Source MAC address
  IPv6: Source MAC address
```

可见，IPv4 和 IPv6 中的数据包，EtherChannel 在默认情况下都是基于源 MAC 地址的负载均衡，而不是基于目的 IP 的负载均衡。

4．查看指定的 EtherChannel 包含的接口

```
SW1#show etherchannel port-channel
                Channel-group listing:
                ----------------------

Group: 1
----------
                Port-channels in the group:
                ---------------------------

Port-channel: Po1
------------

Age of the Port-channel   = 00d:04h:57m:40s
Logical slot/port   = 2/1        Number of ports = 2
GC                  = 0x00000000       HotStandBy port = null
Port state          = Port-channel
```

```
Protocol              =PAGP        //使用的协商协议
Port Security         = Disabled

Ports in the Port-channel:              //下面显示该通道包含的接口有 g0/1 和 g0/2

Index   Load   Port       EC state          No of bits
------+------+------+------------------+-----------
  0     00    Gig0/1     Desirable-Sl       0
  0     00    Gig0/2     Desirable-Sl       0
Time since last port bundled:   00d:04h:50m:04s    Gig0/2
```

实验 26 标准 ACL

一、实验要求

- 理解标准 ACL 的工作方式；
- 掌握标准数字式 ACL 的定义和应用方法；
- 掌握标准命名式 ACL 的定义和应用方法。

二、实验说明

标准 ACL 是通过对网络中的源 IP 地址进行过滤，分为标准数字式 ACL 和标准命名式 ACL。通过本实验的学习，可以为后面掌握更复杂的 ACL 打下基础。

三、实验拓扑

本实验拓扑如图 26-1 所示，要求 PC0 不能 ping 通 PC1，PC2 能 ping 通 PC1，PC0 与 PC2 之间也能相互 ping 通。

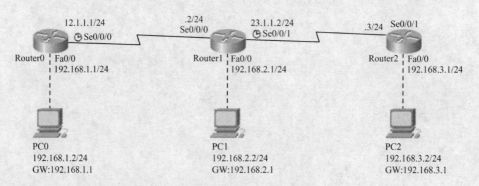

图 26-1 配置标准数字式 ACL 实验拓扑

四、配置过程

（1）在 Router0 上的配置过程

Router0 (config)#int f0/0
Router0 (config-if)#ip add 192.168.1.1 255.255.255.0

```
Router0 (config-if)#no shut
Router0 (config-if)#int s0/0/0
Router0 (config-if)#ip add 12.1.1.1 255.0.0.0
Router0 (config-if)#no shut
Router0 (config-if)#clock rate 1200
Router0 (config-if)#exit
Router0 (config)#router rip
Router0 (config-router)#netw 12.0.0.0
Router0 (config-router)#netw 192.168.1.0
```

（2）在 Router1 上的配置过程

```
Router1(config)#int f0/0
Router1(config-if)#ip add 192.168.2.1 255.255.255.0
Router1(config-if)#no shut
Router1(config-if)#int s0/0/0
Router1(config-if)#ip add 12.1.1.2 255.255.255.0
Router1(config-if)#no shut
Router1(config-if)#int s0/0/1
Router1(config-if)#ip add 23.1.1.1 255.255.255.0
Router1(config-if)#no shut
Router1(config-if)#clock rate 1200
Router1(config-if)#exit
Router1(config)#router rip
Router1(config-router)#netw 12.0.0.0
Router1(config-router)#netw 23.0.0.0
Router1(config-router)#netw 192.168.2.0
```

（3）在 Router2 上的配置过程

```
Router2(config)#int f0/0
Router2(config-if)#ip add 192.168.3.1 255.255.255.0
Router2(config-if)#no shut
Router2(config-if)#int s0/0/1
Router2(config-if)#ip add 23.1.1.3 255.255.255.0
Router2(config-if)#no shut
Router2(config-if)#exit
Router2(config)#router rip
Router2(config-router)#netw 23.0.0.0
Router2(config-router)#netw 192.168.3.0
```

然后配置 PC0、PC1、PC2 的 IP 地址、子网掩码以及网关地址。完成后 PC0 和 PC2 可以 ping 通 PC1，PC0 和 PC2 之间也能互相 ping 通。以上的配置是确保整个网络的基本通信正常的。

然后配置标准数字式 ACL，使 PC0 不能 ping 通 PC1，其余不变。

在路由器 Router1 上，增加以下配置：

```
Router1(config)#access-list 1 deny 192.168.1.2
                //定义 access-list 1，拒绝主机 192.168.1.2 的流量
Router1(config)#access-list 1 permit any
                //定义 access-list 1，允许所有的 IP 地址。这一行不能省，因为在 ACL 最
```

后隐含了一条 deny any 的语句，如果没有这一条，则所有的 IP 地址都将被拒绝，这样 PC2 也不能 ping 通 PC1 了

Router1(config)#int f0/0
Router1(config-if)#ip access-group 1 out
　　　　　　//在 f0/0 的出方向上应用上面定义的 access-list 1。ACL 有进和出两个方向，
　　　　　　//现在是从 PC0 到 PC1 的流量，已经进入了路由器，所以应在出方向上去
　　　　　　//应用所定义的 ACL

五、实验总结

（1）ACL 配置顺序

在配置 ACL 时应注意其配置顺序，先配置的语句在 ACL 语句列表的前面，后配置的在后面，ACL 的工作过程是按 ACL 配置的顺序往后执行，后面的语句只有在前面的语句没有匹配时，才有机会执行，因此 ACL 语句配置的先后次序是非常重要的。

在 Router1 上配置完 ACL 后，在 PC0 上 ping PC1，不能 ping 通，PC2 能 ping 通 PC1，PC0 与 PC2 之间也能相互 ping 通，达到了配置前的要求。

（2）编辑标准数字式 ACL

① 删除 ACL。

Router1(config)#no access-list 1

在删除定义的 ACL 时，只需要删除该 ACL 对应的表号即可，不需要输入具体定义的 ACL 语句。

② 取消 ACL 在接口上的应用。

Router1(config)#int f0/0
Router1(config-if)#no ip access-group 1 out

③ 编辑 ACL。

对数字式 ACL，在编辑 ACL 时，不能删除一行，也不能在原 ACL 语句中插入一行。较快的编辑方法是：先用 show run 显示当前配置，然后把所配的 ACL 语句复制下来放到记事本中进行编辑，再在路由器上删除原 ACL，最后把记事本中编辑好的 ACL 语句复制到路由器中。

这里需要注意的是：编辑好的 ACL 语句在复制到路由器前，一定要先删除原 ACL，否则只会在原 ACL 语句后添加新的语句。

（3）标准数字式 ACL 的放置位置

所谓放置位置，是指将所定义的 ACL 放到哪个路由器、哪个接口上，以及方向如何指定。其放置规则是：标准 ACL（包括数字式和命名式），放到距目标最近的那个接口上，并且方向一般用外出方向"out"。如本例中，就是放到 PC1 最近的路由器接口 f0/0 上，方向用"out"，其理由是：标准 ACL 仅对源 IP 进行过滤，如果放置距源 IP 更近的地方，那么将会影响使 ACL 过滤掉到达其他目标的流量，如本例中，如果将定义的 ACL 放置到路由器 Router0 的端口 f0/0 或 s0/0/0，或放到路由器 Router1 的 s0/0/0 端口上，将会使 PC0 除了无法 ping 通 PC1 外，还无法 ping 通 PC2。

本例中，Router1 的 f0/0 接口直接与目标 PC1 相连，是距目标最近的接口，此时的数据流量是从路由器出来，所以方向是 out，而不是 in（如果是在 Router1 的 s0/0/0 上应用 ACL，此时数据流量还没有进入路由器，方向应该用 in）。

（4）标准命名式 ACL

由于使用数字式 ACL 在编辑时不方便，用数字编号不直观，并且当网络中需要配置超过 99 个 ACL 时，数字式 ACL 不能实现，而命名式 ACL 则没有这些缺点。

下面用标准命名式 ACL 来完成上面标准数字式 ACL 同样的要求，接着上面的实验配置，在路由器 Router1 上配置标准命名式 ACL：

```
Router1(config)#no access-list 1                        //先删除前面配置的ACL
Router1(config)#int f0/0
Router1(config-if)#no ip access-group 1 out             //取消前面ACL在f0/0上的应用
Router1(config)#ip access-list standard denypc0         //创建命名式ACL denypc0
Router1(config-std-nacl)# deny host 192.168.1.2         //拒绝来自主机192.168.1.2的流量
Router1(config-std-nacl)#permit any                     //许可其余流量
Router1(config-std-nacl)#exit
Router1(config)#int f0/0
Router1(config-if)#ip access-group denypc0 out          //将命名式ACLdenypc0应用在f0/0接口的出
                                                        //方向上
```

配置完成后测试各 PC 间的连通性，其结果与标准数字式 ACL 相同。

标准命名式 ACL 的编辑比标准数字式 ACL 更方便，可以在不删除 ACL 的情况下完成。

例如，现需要使 PC0 能 ping 通 PC1，使 PC2 不能 ping 通 PC1，可以 Router1 上进行如下配置：

```
Router1(config)#ip access-list standard denypc0
Router1(config-std-nacl)#no deny 192.168.1.2            //可直接删除命名式ACL中的一行
Router1(config-std-nacl)#no permit any                  //这条也应删除，否则，它将在其他ACL语句之前，
                                                        //使后面的ACL语句不起作用
Router1(config-std-nacl)#deny 192.168.3.2               //配上拒绝PC2的ACL
Router1(config-std-nacl)#permit any                     //在最后重新配上许可其余流量
Router1(config-std-nacl)#exit
```

对整个标准命名式 ACL 的删除和取消在接口上的应用，与标准数字式 ACL 方法相同。

实验 27 扩展 ACL

一、实验要求

- 理解扩展 ACL 的工作方式；
- 掌握扩展数字式 ACL 的定义和应用方法；
- 掌握扩展命名式 ACL 的定义和应用方法。

二、实验说明

标准 ACL 只能对源地址进行过滤，而扩展 ACL 则可以同时对源地址、目标地址、协议、端口号、时间范围等来过滤，可见扩展 ACL 的功能比标准 ACL 更强。

在上一个实验的基础上，继续配置扩展数字式 ACL。以图 27-1 为拓扑图：禁止 192.168.1.0/24 网络中 PC3 到 192.168.3.0/24 网络的 FTP 服务，PC4 到 192.168.3.0/24 网络的 Telnet 服务，192.168.1.0/24 网络的所有主机允许使用到 192.168.3.0/24 网络的 WWW 服务，整个 192.168.1.0/24 网络不允许 ping 通 192.168.3.0/24 网络的主机，禁止路由器 Router0 ping 通路由器 Router2 的 s0/0/1 端口。

三、实验拓扑

本实验的拓扑是在图 26-1 的基础上增加一台交换机 SW0 和两台计算机 PC3、PC4，如图 27-1 所示。

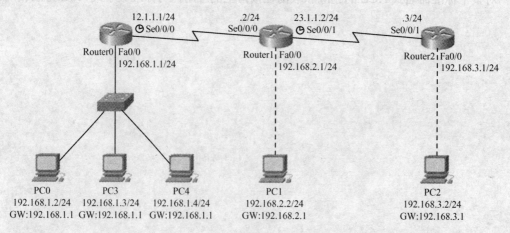

图 27-1 扩展数字式 ACL 实验拓扑

四、实验过程及实验总结

① 沿用前一个实验对路由器的基本配置，需删除上一个实验在路由器 Router1 上的原 ACL 配置：

  Router1(config)#no access-list standard denypc0
  Router1(config)#int f0/0
  Router1(config-if)#no ip access-group denypc0 out

然后在路由器 Router0 上的配置新的 ACL：

  Router0(config)#access-list 100 deny tcp host 192.168.1.3 192.168.3.0 0.0.0.255 eq 21
          //eq 21 也可写为 eq ftp
  Router0(config)#access-list 100 deny tcp host 192.168.1.4 192.168.3.0 0.0.0.255 eq 23
          //eq 23 也可写为 eq telnet
  Router0(config)#access-list 100 permit tcp 192.168.1.0 0.0.0.255 192.168.3.0 0.0.0.255 eq www
          //eq www 也可写为 eq 80
  Router0(config)#access-list 100 deny icmp 192.168.1.0 0.0.0.255 192.168.3.0 0.0.0.255
  Router0(config)#access-list 100 permit ip any any
  Router0(config)#int f0/0
  Router0(config-if)#ip access-group 100 in
  Router0(config-if)#exit

在路由器 Router1 上新增的配置：

  Router1(config)#access-list 101 deny icmp host 12.1.1.1 host 23.1.1.3
  Router1(config)#access-list 101 permit ip any any
  Router1(config)#int s0/0/0
  Router1(config-if)#ip access-group 101 in
          //禁止 Router0 ping Router2 的 s0/0/1 端口，不能在 Router0 上配置 ACL 来限制，只能在下一跳路由器 Router1 入口 s0/0/0 上作此限制

这里有一点需要特别注意：如果没有配置"access-list 101 permit ip any any"，ACL 的末尾有一条默认拒绝所有流量的语句，将会拒绝包括路由协议在内的所有流量。在本例中保证基本通信使用的是 RIP 路由协议，这将会导致路由器 Router1 不能通过 RIP 学到 192.168.1.0 网络（当然 Router2 也不能学到 192.168.1.0 网络）。在 RIP 的更新超过 240s 后，再查看 Router1 的路由表：

  Router1#show ip route
  ……
    12.0.0.0/24 is subnetted, 1 subnets
  C 12.1.1.0 is directly connected, Serial0/0/0
    23.0.0.0/24 is subnetted, 1 subnets
  C 23.1.1.0 is directly connected, Serial0/0/1
  C 192.168.2.0/24 is directly connected, FastEthernet0/0
  R 192.168.3.0/24 [120/1] via 23.1.1.3, 00:00:10, Serial0/0/1

可以发现，已经没有了到 192.168.1.0 网络的路由条目（同样在 Router2 的路由表中也没有），此时，凡是经过路由器 Router1 的 S0/0/0 接口的任何通信（包括 ping、telnet、ftp、www 等）都将失败，这是因为路由协议没正常学习，连基本的网络通信都没有了。

153

无论是基本 ACL，还是扩展 ACL，在配置时，一定要注意考虑到在 ACL 的后面还有一条默认拒绝所有流量的语句。

② 扩展 ACL 放置的位置。

由于扩展 ACL 对数据包的限制非常"准确"，因此可以将扩展 ACL 应用在距源数据包更近的地方，以减少在网络中不必要的流量。例如，在 R0 上配置的 ACL，是将 ACL 放到了 Router 0 的 F0/0 接口上。如果是放到 Router 0 的 S0/0/0 接口上，则那些被拒绝发往目标 IP 的数据包都需要先进入路由器，路由器根据协议进行计算，得出的结果需要从 S0/0/0 端口发出去，但到了 S0/0/0 端口后，其结果仍然被拒绝掉，这样就浪费了路由器的资源。

③ 扩展命名式 ACL。

扩展命名式 ACL 的编辑方法类似于标准命名式 ACL，其功能与扩展数字式 ACL 类似，现以图 27-1 为例，将配置在路由器 Router0 上的扩展数字式 ACL 改配置为扩展命名式 ACL。

首先删除掉路由器 Router0 上的原 ACL 配置：

  Router0(config)#no access-list 100
  Router0(config)#int f0/0
  Router0(config-if)#no ip access-group 100 in

在路由器 Router0 上配置扩展命名式 ACL 的过程：

  Router0(config)#ip access-list extended EXT-ACL
  Router0(config-ext-nacl)#deny tcp host 192.168.1.3 192.168.3.0 0.0.0.255 eq 21
  Router0(config-ext-nacl)#deny tcp host 192.168.1.4 192.168.3.0 0.0.0.255 eq 23
  Router0(config-ext-nacl)#permit tcp 192.168.1.0 0.0.0.255 192.168.3.0 0.0.0.255 eq www
  Router0(config-ext-nacl)#deny icmp 192.168.1.0 0.0.0.255 192.168.3.0 0.0.0.255
  Router0(config-ext-nacl)#permit ip any any
  Router0(config-ext-nacl)#exit
  Router0(config)#int f0/0
  Router0(config-if)#ip access-group EXT-ACL in

实验 28 基于时间的 ACL

一、实验要求

- 了解基于时间 ACL 的作用；
- 掌握基于时间的 ACL 的配置方法；
- 掌握基于时间的 ACL 的调试方法。

二、实验说明

基于时间的 ACL 是在普通 ACL 的基础上加入了对时间范围的控制功能。通过基于时间的 ACL，可使企业对员工在指定时间内（如工作时间内）不能利用公司网络任意访问网络，限制员工只能利用网络做与工作相关等工作。

三、实验拓扑

本实验的拓扑图如图 28-1 所示。

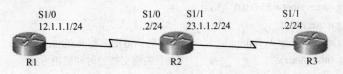

图 28-1 基于时间的 ACL 实验拓扑

四、实验过程

（1）R1 的配置过程

```
Router>en
Router#conf t
Router(config)#host R1
R1(config)#line console 0
R1(config-line)#exec-timeout 0 0          //确保 console 线路永不超时
R1(config-line)#exit
R1(config)#int s1/0
R1(config-if)#ip add 12.1.1.1 255.255.255.0
R1(config-if)#no shut
```

```
R1(config-line)#clock rate 1200
R1(config-if)#exit
R1(config)#router rip                //配置 RIP 协议的目的是全网能正常通信
R1(config-router)#netw 12.0.0.0
R1(config-router)#ver 2
R1(config-router)#exit
```

(2) R2 的配置过程

```
Router>en
Router#conf t
Router(config)#host R2
R2(config)#line console 0
R2(config-line)#exec-timeout 0 0
R2(config-line)#exit
R2(config)#int s1/0
R2(config-if)#ip add 12.1.1.2 255.255.255.0
R2(config-if)#no shut
R2(config-if)#clock rate 1200
R2(config-if)#int s1/1
R2(config-if)#ip add 23.1.1.2 255.255.255.0
R2(config-if)#no shut
R2(config-if)#exit
R2(config)#router rip
R2(config-router)#ver 2
R2(config-router)#no auto-summary
R2(config-router)#netw 12.0.0.0
R2(config-router)#netw 23.0.0.0
R2(config router)#exit
R2(config)#time-range worktime       //定义时间范围名称 worktime
R2(config-time-range)#?              //查看当前支持的时间范围方式
Time range configuration commands:
  absolute    absolute time and date
                                     //指定一个绝对的时间范围,如设定时间段为 2015 年 4 月
                                     //1 日的 0:00～2015 年 5 月 1 日的 0:00,其格式是
                                     //absolute start 0:00 1 April 2015 end 0:00 1 may 2015
  default     Set a command to its defaults
  exit        Exit from time-range configuration mode
  no          Negate a command or set its defaults
  periodic    periodic time and date
                                     //指定重复发生在一个星期内的时间范围,可以与 absolute 套
                                     //用,表示在一个时期内的一星期内重复发生
```

下面查看其后续命令:

```
R2(config-time-range)#periodic ?
  Friday      Friday                 //指明一周的哪一天
  Monday      Monday
  Saturday    Saturday
```

```
                Sunday      Sunday
                Thursday    Thursday
                Tuesday     Tuesday
                Wednesday   Wednesday
                daily       Every day of the week      //一星期的每一天
                weekdays    Monday thru Friday         //周内的工作日,即星期一至星期五
                weekend     Saturday and Sunday        //周末,即星期六、星期天
    R2(config-time-range)#periodic weekdays 8:00 to 12:00
                            //这里定义工作日的上午 8～12 点
    R2(config-time-range)#periodic weekdays 14:00 to 18:00
                            //这里定义工作日的下午 14～18 点
    R2(config-time-range)#exit
    R2(config)#access-list 100 permit tcp any any eq 53       //允许 DNS 解析
    R2(config)#access-list 100 permit tcp any any eq smtp     //允许发邮件
    R2(config)#access-list 100 permit tcp any any eq pop3     //允许收邮件
                            //上面这几条是在任何时间都允许的,包括工作时间
    R2(config)#access-list 100 deny ip any any time-range worktime
                            //在工作时间内,拒绝所有的 IP 流量。注意,这里的 IP 流量包括了所有的网络流
                            //量,如 TCP、ICMP、UDP、FTP、Telnet、WWW 等,当然,根据 ACL 过滤
                            //规则,在前面已经许可的流量是允许通过的
    R2(config)#access-list 100 permit ip any any
                            //允许所有 IP 流量
    R2(config)#int s1/0
    R2(config-if)#ip access-group 100 in    //将 ACL 放置在 R2 的 s1/0 端口上
    R2(config-if)#end
```

（3）R3 的配置过程
```
    Router>en
    Router#conf t
    Router(config)#host R3
    R3(config)#line console 0
    R3(config-line)#exec-timeout 0 0
    R3(config-line)#exit
    R3(config)#int s1/1
    R3(config-if)#ip add 23.1.1.3 255.255.255.0
    R3(config-if)#no shut
    R3(config-if)#exit
    R3(config)#router rip
    R3(config-router)#ver 2
    R3(config-router)#no auto-summary
    R3(config-router)#netw 23.0.0.0
    R3(config-router)#exit
    R3(config)#enable password ccna
    R3(config)#line vty 0 4           //在 R3 上配置虚拟线路,用于 telnet 测试
    R3(config-line)#password ccnp
    R3(config-line)#login
    R3(config-line)#exit
```

五、实验总结

在上述配置完成后，如果在"worktime"所定义的工作时间段内，即周一到周五的 8：00～12：00，14：00～18：00 内，可以进行收发邮件工作，其他操作不能成功，例如在 R1 上，对 R3 进行 telnet，是不能成功的，而其他时间内可以，下面是测试过程：

（现在的时间是星期五的 14：40）

在 R1 上 telnet R3：

    R1#telnet 23.1.1.3
    Trying 23.1.1.3 ...
    % Destination unreachable; gateway or host down

可见，telnet 被拒绝。

然后，到 R2 上查看 ACL：

    R2#show ip access-lists 100
    Extended IP access list 100
        10 permit tcp any any eq domain
        20 permit tcp any any eq smtp
        30 permit tcp any any eq pop3
        40 deny ip any any time-range worktime **(active)** (17 matches)
        50 permit ip any any

可见，现在的 worktime 处于 active（活跃）状态。

现在把 R2 中定义的 worktime 时间进行修改：

    R2(config)#time-range worktime
    R2(config-time-range)#no periodic weekdays 14:00 to 18:00
    R2(config-time-range)#periodic weekdays 15:00 to 18:00

把下午的时间改为了 15:00 ～18:00 后，查看一下修改效果：

    R2#show time-range
    time-range entry: worktime (inactive)
        periodic weekdays 8:00 to 12:00
        periodic weekdays 15:00 to 18:00   //已改为了 15:00 ～18:00
        used in: IP ACL entry

现在再从 R1 上去 telnet R3：

    R1#telnet 23.1.1.3
    Trying 23.1.1.3 ... Open

    User Access Verification

    Password:
    Password:
    R3>en
    Password:
    R3#

此时，R1 能正常 telnet 到 R3 上。

实验 29 反射 ACL

一、实验要求

- 了解反射 ACL 的作用；
- 理解 "established" 方式与反射 ACL 的区别；
- 掌握反射 ACL 的配置方法。

二、实验说明

① 反射 ACL 不能在编号 ACL 或标准命名式 ACL 中定义，只能在扩展命名 ACL 中定义。

② 在扩展 ACL 中，有一个 "established"（已建立）的参数，此参数只能对 TCP 连接起作用，如 telnet、FTP、HTTP 等，这些 TCP 流量建立后，外网返回的数据包中含有 ACK 参数，但对 UDP、ICMP 不起作用，因为它们传输数据时不建立连接，没有 ACK 参数。而反射 ACL 则不同，它通过内网在访问外网时，主动生成一个新的临时条目，此条目既可以通过检查数据包中的 ACL 或 RST 位来确认是否为内网向外网发起的连接，不可以通过检查源和目的端口号来确认，所以反射 ACL 对所有的网络流量都能起作用。

③ 反射 ACL 只允许内网向外网主动产生的流量，而阻止从外网主动向内网产生的流量，这样可以起到防火墙的作用。

④ 反射 ACL 是在有流量产生时临时自动产生，在会话结束后就自动删除。

⑤ 反射 ACL 是嵌套在一个扩展命名 ACL 下的。

三、实验拓扑

本实验的拓扑图如图 29-1 所示。

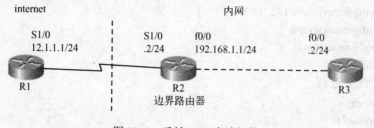

图 29-1 反射 ACL 实验拓扑

四、实验过程

(1) R1 的配置过程

Router>
Router>en
Router#conf t
Router(config)#host R1
R1(config)#int s1/0
R1(config-if)#ip add 12.1.1.1 255.255.255.0
R1(config-if)#no shut
R1(config-if)#clock rate 1200
R1(config-if)#exit
R1(config)#router rip
R1(config-router)#netw 12.1.1.0
R1(config-router)#ver 2
R1(config-router)#exit
R1(config)#enable password ccna
R1(config)#line vty 0 4   //配置虚拟线路，许可 telnet
R1(config-line)#password ccnp
R1(config-line)#login
R1(config-line)#exit
R1(config)#

(2) R2 的配置过程

Router>en
Router#conf t
Router(config)#host R2
R2(config)#int s1/0
R2(config-if)#ip add 12.1.1.2 255.255.255.0
R2(config-if)#no shut
R2(config-if)#int f0/0
R2(config-if)#ip add 192.168.1.1 255.255.255.0
R2(config-if)#no shut
R2(config-if)#exit
R2(config)#router rip
R2(config-router)#netw 12.1.1.0
R2(config-router)#netw 192.168.1.0
R2(config-router)#ver 2
R2(config-router)#exit
R2(config)#

(3) R3 的配置过程

Router>en
Router#conf t
Router(config)#host R3
R3(config)#int f0/0

R3(config-if)#ip add 192.168.1.2 255.255.255.0
R3(config-if)#no shut
R3(config-if)#exit
R3(config)#router rip
R3(config-router)#netw 192.168.1.0
R3(config-router)#exit
R3(config)#enable password ccna
R3(config)#line vty 0 4
R3(config-line)#password ccnp
R3(config-line)#login
R3(config-line)#exit
R3(config)#

五、实验总结

① 在 R1 上对 R3 的 ping 和 telnet 测试：

R1#ping 192.168.1.2

Type escape sequence to abort.
Sending 5, 100-byte ICMP Echos to 192.168.1.2, timeout is 2 seconds:
!!!!!
Success rate is 100 percent (5/5), round-trip min/avg/max = 16/37/68 ms
R1#telnet 192.168.1.2
Trying 192.168.1.2 ... Open

User Access Verification

Password:
R3>en
Password:
R3#exit

[Connection to 192.168.1.2 closed by foreign host]
R1#

② 在 R3 上对 R1 的 ping 和 telnet 测试：

R3#ping 12.1.1.1

Type escape sequence to abort.
Sending 5, 100-byte ICMP Echos to 12.1.1.1, timeout is 2 seconds:
!!!!!
Success rate is 100 percent (5/5), round-trip min/avg/max = 12/24/72 ms
R3#telnet 12.1.1.1
Trying 12.1.1.1 ... Open

User Access Verification

Password:
R1>en
Password:
R1#exit

[Connection to 12.1.1.1 closed by foreign host]
R3#

可见，现在 R1 和 R3 之间的 ping 和 telnet 都能正常进行。

③ 然后，在 R2 上增加自反 ACL 的配置：

R2(config)#ip access-list extended outacl   //建立扩展 ACL，取名 outacl
R2(config-ext-nacl)#permit ip any any reflect outip
                    //许可所有 IP 流量，并对外出 IP 流量进行反射，将反射 ACL 嵌套在扩展
                    //ACL 中，取名为 outip
R2(config-ext-nacl)#exit
R2(config)#ip access-list extended inacl   //建立扩展 ACL，取名 inacl
R2(config-ext-nacl)#evaluate outip   //调用前面创建的反射列表，称为评估反射列表
R2(config-ext-nacl)#int s1/0
R2(config-if)#ip access-group outacl out   //调用 ACL，在外出方向上做反射
R2(config-if)#ip access-group inacl in   //调用 ACL，在进入方向上做评估

④ 在 R2 上增加自反 ACL 的配置后，再一次在 R1 上对 R3 的 ping 和 telnet 进行测试：

R1#ping 192.168.1.2

Type escape sequence to abort.
Sending 5, 100-byte ICMP Echos to 192.168.1.2, timeout is 2 seconds:
U.U.U
Success rate is 0 percent (0/5)
R1#telnet 192.168.1.2
Trying 192.168.1.2 ...
% Destination unreachable; gateway or host down

可见，R1 不能 ping 通 R3，也不能 telnet 到 R3。

⑤ 再在 R3 上对 R1 的 ping 和 telnet 测试：

R3#ping 12.1.1.1

Type escape sequence to abort.
Sending 5, 100-byte ICMP Echos to 12.1.1.1, timeout is 2 seconds:
!!!!!
Success rate is 100 percent (5/5), round-trip min/avg/max = 12/61/132 ms
R3#telnet 12.1.1.1
Trying 12.1.1.1 ... Open

User Access Verification

Password:
R1>en
Password:
R1#exit

[Connection to 12.1.1.1 closed by foreign host]
R3#

可见，R3 能 ping 通 R1，也能 telnet 到 R1。

⑥ 在 R3 telnet 到 R1 后，在 R2 上查看反射控制列表：

R2#show ip access-lists
Extended ip access list inacl
 10 evaluate outip
Extended ip access list outacl
 10 permit ip any any reflect outip (125 matches)
Reflexive ip access list outip
 permit tcp host 12.1.1.1 eq telnet host 192.168.1.2 eq 11002 (83 matches) (time left 285)

其中，(time left 285)表示该反射条目还有 285s 被删除（默认是 300s），这个条目被删除的时间可以修改：

R2(config)#ip reflexive-list timeout ?
 <1-2147483> timeout in seconds // <1-2147483>是秒钟时间范围

R2(config)#ip reflexive-list timeout 10 //改为 10s 后删除

⑦ 过了删除期后，再查看 R2 上的反射控制列表：

R2#show ip access-lists
Extended IP access list inacl
 10 evaluate outip
Extended IP access list outacl
 10 permit ip any any reflect outip (125 matches)
Reflexive IP access list outip

可见，反射控制列表已自动删除了。

实验 30 路由器 DHCP 的配置

一、实验要求

- 掌握 DHCP 服务的基本配置；
- 掌握 DHCP 中继的配置；
- 掌握 DHCP 客户端的设置。

二、实验说明

在复杂的具有层次的网络中，配有 DHCP 服务的路由器一般放置在网络中心，而 DHCP 客户端和 DHCP 服务器可能不在同一个网段中，这就需要使用 DHCP 中继服务功能，为 DHCP 客户端自动分配 IP 地址。

三、实验拓扑

本实验的拓扑图如图 30-1 所示。

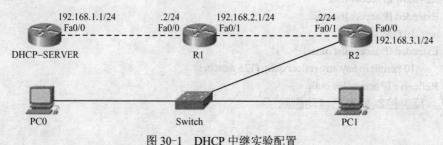

图 30-1 DHCP 中继实验配置

四、配置过程

（1）在路由器 DHCP-SERVER 上的配置过程

DHCP-SERVER(config)#int f0/0
DHCP-SERVER(config-if)#ip add 192.168.1.1 255.255.255.0
DHCP-SERVER(config-if)#no shut
DHCP-SERVER(config-if)#exit
DHCP-SERVER(config)#router rip //配置 RIP 协议，使网络保证基本连通性
DHCP-SERVER(config-router)#netw 192.168.1.0
DHCP-SERVER(config-router)#exit
DHCP-SERVER(config)#ip dhcp pool pool1

　　　　　　　　　　　　　　　　//配置 DHCP 地址池,命名 pool1
　　　　DHCP-SERVER(dhcp-config)#network 192.168.3.0 255.255.255.0
　　　　　　　　　　　　　　　　//配置 DHCP 分配的网段,此网段是 DHCP 客户端的网关所在的
　　　　　　　　　　　　　　　　//网段,而非本路由器端口所在网段
　　　　DHCP-SERVER(dhcp-config)#default-router 192.168.3.1
　　　　　　　　　　　　　　　　//配置 DHCP 客户端的网关地址,非本路由器接口地址
　　　　DHCP-SERVER(dhcp-config)#dns-server 61.128.128.68
　　　　DHCP-SERVER(dhcp-config)#lease 8　　//设置 IP 地址的租期为 8 天
　　　　DHCP-SERVER(dhcp-config)#exit
　　　　DHCP-SERVER(config)#ip dhcp excluded-address 192.168.3.1
　(2) 路由器 R1 上的配置过程
　　　　R1(config)#int f0/0
　　　　R1(config-if)#ip add 192.168.1.1 255.255.255.0
　　　　R1(config-if)#no shut
　　　　R1(config-if)#int f0/1
　　　　R1(config-if)#ip add 192.168.2.1 255.255.255.0
　　　　R1(config-if)#no shut
　　　　R1(config-if)#exit
　　　　R1(config)#router rip
　　　　R1(config-router)#netw 192.168.1.0
　　　　R1(config-router)#netw 192.168.2.0
　(3) 路由器 R2 上的配置过程
　　　　R2(config)#int f0/1
　　　　R2(config-if)#ip add 192.168.2.2 255.255.255.0
　　　　R2(config-if)#no shut
　　　　R2(config-if)#int f0/0
　　　　R2(config-if)#ip add 192.168.3.1 255.255.255.0
　　　　R2(config-if)#no shut
　　　　R2(config-if)#exit
　　　　R2(config)#router rip
　　　　R2(config-router)#netw 192.168.2.0
　　　　R2(config-router)#netw 192.168.3.0
　　　　R2(config-router)#exit
　　　　R2(config)#int f0/0　　　　　　　　//在距 DHCP 客户端最近的路由器接口,即网关接口上配置
　　　　　　　　　　　　　　　　　　　　　　//DHCP 中继功能
　　　　R2(config-if)#ip helper-address 192.168.1.1
　　　　　　　　　　　　　　　　　　　　　　//配置帮助地址,此地址是 DHCP 服务器地址,即把网关接口收
　　　　　　　　　　　　　　　　　　　　　　// 到的 DHCP discover 请求广播包转发到 DHCP 服务器
　　　　　　　　　　　　　　　　　　　　　　//192.168.1.1 上
　　在此,交换机 Switch 不需要配置,仅是为了起扩充端口的作用而已。

五、实验总结

　(1) 在 Cisco Packet Tracer 实验中的 PC 上设置
　　以 PC0 为例,在 Cisco Packet Tracer 上,按图 30-2 所示,设置为 DHCP 自动获取 IP 方式即可。

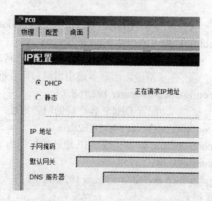

图 30-2 选择 DHCP 方式

然后，如图 30-3 所示，PC0 和 PC1 已由 DHCP 服务器通过中继方式成功为 PC 分配了 IP 地址。

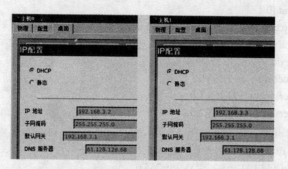

图 30-3 DHCP 通过中继方式自动分配 IP

一般来说，DHCP 服务器是从没有分配使用的、最小的 IP 地址开始自动为 DHCP 客户端分配 IP 地址。

（2）在真实计算机上的设置

如果是在真实的计算机上，可按图 30-4 所示设置自动获得 IP 地址及 DNS 等信息，然后在真实的 PC 上，在 CMD 模式下，执行"ipconfig/all"命令，如图 30-5 所示，可以查看到 PC 上已自动获得的相关地址信息。

图 30-4 配置真实 PC 自动获得 IP 地址

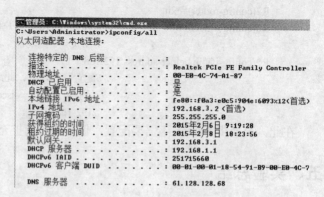

图 30-5 查看真实 PC 上自动获得的 IP 地址

实验 31　IPv6 静态路由

一、实验要求

- 掌握如何启动 IPv6 路由；
- 掌握如何配置 IPv6 地址；
- 掌握 IPv6 静态路由的配置方法。

二、实验说明

IPv6 的静态路由工作原理与 IPv4 相同，通过本实验学习在路由器上如何配置 IPv6，为后面配置动态路由和 IPv6 隧道打下基础，更详细的实验说明在下一个实验"IPv6 RIPng"中讲述。

三、实验拓扑

IPv6 静态路由的配置使用的拓扑图如图 31-1 所示。

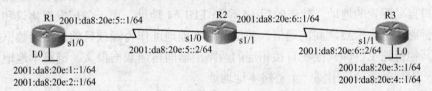

图 31-1　IPv6 静态路由配置实验拓扑

四、配置过程

(1) R1 上的配置过程

```
R1(config)#
R1(config)#ipv6 unicast-routing          //启用 IPv6 路由，路由器默认不支持 IPv6
R1(config)#int s1/0
R1(config-if)#ipv6 address 2001:da8:20e:5::1/64   //配置 IPv6 地址
R1(config-if)#no shut
R1(config-if)#int L0
R1(config-if)#ipv6 add 2001:da8:20e:1::1/64       //配置 IPv6 地址
R1(config-if)#ipv6 add 2001:da8:20e:2::1/64       //配置第二个 IPv6 地址，一个接口可配置多个
                                                  //IPv6 地址
```

```
R1(config-if)#exit
R1(config)#ipv6 route ::/0 2001:da8:20e:5::2        //配置 IPv6 静态路由，注意这里的::/0 表示
                                                    //0:0:0:0:0:0:0:0/0
```

在接口下配置 IPv6 地址时，除了使用上面的手工指定方式外，还可以使用"EUI-64"方式自动生成 IPv6 地址，例如，在配置上面 s1/0 的 IPv6 地址时可这样：

```
R1(config)#int s1/0
R1(config-if)#ipv6 address 2001:da8:20e:5::/64 eui-64
```

EUI-64 地址是由在 48 位 MAC 地址中的 24 位之后插入 16 位 0xFFFE，从而形成唯一的 64 位接口标识符。

配置完毕后，可以查看在接口上配置的 IPv6 地址：

```
R1#show ipv6 interface brief
FastEthernet0/0            [administratively down/down]
    unassigned
Serial1/0                  [up/down]
    FE80::C800:BFF:FE08:0
    2001:DA8:20E:5::1
Serial1/1                  [administratively down/down]
    unassigned
Loopback0                  [up/up]
    FE80::C800:BFF:FE08:0
    2001:DA8:20E:1::1
    2001:DA8:20E:2::1
```

其中 FE80::C800:BFF:FE08:0 是 IPv6 的私有地址，称为链路本地地址，其前三个字符可以是 FE8、FE9、FEA、FEB，其中最常见的是以 FE80 开头的链路本地地址。链路本地地址是直连接口自动配置的地址，其中的后 64 位是 EUI-64 地址，用于邻居发现协议和无状态自动配置中链路本地上节点之间的通信。使用链路本地地址作为源或目的地址的数据报文不会被转发到其他链路上，即只能在直接相连的两路由器间传递数据报文。链路本地地址对节点来说是唯一的，多个接口用同一个链路本地地址。

还有一种 IPv6 的私有地址，称为站点本地地址，其前三个字符可以是 FEC、FED、FEE、FEF。站点本地地址与 IPv4 中的私有地址类似。使用站点本地地址作为源或目的地址的数据报文不会被转发到本站点（相当于一个私有网络）外的其他站点。

（2）R2 上的配置过程

```
R2(config)#ipv6 unicast-routing
R2(config)#int s1/0
R2(config-if)#ipv6 address 2001:da8:20e:5::2/64
R2(config-if)#no shut
R2(config-if)#int s1/1
R2(config-if)#ipv6 address 2001:da8:20e:6::1/64
R2(config-if)#no shut
R2(config-if)#exit
R2(config)#ipv6 route 2001:da8:20e:1::/64 2001:da8:20e:5::1
R2(config)#ipv6 route 2001:da8:20e:2::/64 2001:da8:20e:5::1
R2(config)#ipv6 route 2001:da8:20e:3::/64 2001:da8:20e:6::2
```

R2(config)#ipv6 route 2001:da8:20e:4::/64 2001:da8:20e:6::2
//配置四条 IPv6 静态路由，分别到达四个不同的网络（地址），R1 和 R3 上
//的 Loopback 接口各需两条

(3) R3 上的配置过程

R3(config)#ipv6 unicast-routing
R3(config)#int s1/1
R3(config-if)#ipv6 address 2001:da8:20e:6::2/64
R3(config-if)#no shut
R3(config-if)#int L0
R3(config-if)#ipv6 address 2001:da8:20e:3::1/64
R3(config-if)#ipv6 address 2001:da8:20e:4::1/64
R3(config-if)#no shut
R3(config-if)#exit
R3(config)#ipv6 route ::/0 2001:da8:20e:6::1

配置完成后，在各路上器上互 ping 所配置的所有 IPv6 地址，均可 ping 通。

实验 32　IPv6 RIPng

一、实验要求

- 了解 IPv6 RIPng 的配置方法；
- 掌握查看 IPv6 接口信息的方法；
- 掌握查看 IPv6 路由表的方法；
- 掌握查看 IPv6 RIPng 数据库的方法；
- 掌握查看 RIPng 的相关信息的方法。

二、实验说明

IPv6 RIPng 与 IPv4 RIPv2 的区别：IPv4 RIPv2 数据包更新使用的组播地址是 224.0.0.9，而 RIPng 使用的组播地址是 FF02::9；在配置 RIPng 的时候，IPv4 公告网络是在路由进程中进行的，而 RIPng 采用先配置进程，然后在相应的接口下公告网络。

三、实验拓扑

IPv6 RIPng 实验拓扑如图 32-1 所示。

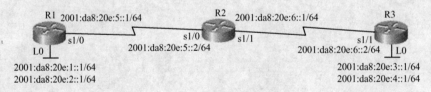

图 32-1　IPv6 RIPng

四、实验过程

（1）R1 的配置过程

```
Router>en
Router#conf t
Router(config)#line console 0
Router(config-line)#exec-timeout 0 0
```

```
Router(config-line)#exit
Router(config)#host R1
R1(config)#ipv6 unicast-routing      //路由器默认不支持 IPv6 功能,在此开启
R1(config)#int s1/0
R1(config-if)#ipv6 address 2001:da8:20e:5::1/64    //配置 IPv6 地址
R1(config-if)#no shut
R1(config-if)#int L0
R1(config-if)#ipv6 address 2001:da8:20e:1::1/64
R1(config-if)#ipv6 address 2001:da8:20e:2::1/64
R1(config-if)#exit
R1(config)#ipv6 router rip AAA
              //启用 RIPng 协议,进程号取名为 AAA,进程号只有本地意义
R1(config-rtr)#int L0
R1(config-if)#ipv6 rip AAA enable
              //在 L0 上启用 RIPng 进程 AAA,将宣告 L0 接口运行 RIPng 协议,这与 RIP 协
              //议不同,在 RIP 协议中,是在 RIP 路由进程中宣告网络号
R1(config-if)#int s1/0
R1(config-if)#ipv6 rip AAA enable    //在 s1/0 端口上启用 RIPng 进程 AAA
R1(config-if)#end
```
(2) R2 的配置过程
```
Router>en
Router#conf t
Router(config)#lin
Router(config)#line console 0
Router(config-line)#exec-timeout 0 0
Router(config-line)#exit
Router(config)#host R2
R2(config)#ipv6 unicast-routing
R2(config)#int s1/0
R2(config-if)#ipv6 address 2001:da8:20e:5::2/64
R2(config-if)#no shut
R2(config-if)#int s1/1
R2(config-if)#ipv6 address 2001:da8:20e:6::1/64
R2(config-if)#no shut
R2(config-if)#exit
R2(config)#ipv6 router rip BBB
              //进程号取名为 AAA,进程号只有本地意义,所以可以与前面的 R1 进
              //程号不一样
R2(config-rtr)#int s1/0
R2(config-if)#ipv6 rip BBB enable
R2(config-if)#int s1/1
R2(config-if)#ipv6 rip BBB enable
R2(config-if)#end
```
(3) R3 的配置过程
```
Router>en
Router#conf t
```

```
Router(config)#line console 0
Router(config-line)#exec-timeout 0 0
Router(config-line)#exit
Router(config)#host R3
R3(config)#ipv6 unicast-routing
R3(config)#int s1/1
R3(config-if)#ipv6 address 2001:da8:20e:6::2/64
R3(config-if)#no shut
R3(config-if)#int L0
R3(config-if)#ipv6 address 2001:da8:20e:3::1/64
R3(config-if)#ipv6 address 2001:da8:20e:4::1/64
R3(config-if)#no shut
R3(config-if)#exit
R3(config)#ipv6 router rip CCC
R3(config-rtr)#int s1/1
R3(config-if)#ipv6 rip CCC enable
R3(config-if)#int L0
R3(config-if)#ipv6 rip CCC enable
R3(config-if)#end
R3#
```

五、实验总结

（1）连通性测试

分别在 R1 和 R3 上 ping 对方路由器，来验证网络的连通性：

```
R1#ping ipv6 2001:da8:20e:6::2

Type escape sequence to abort.
Sending 5, 100-byte ICMP Echos to 2001:DA8:20E:6::2, timeout is 2 seconds:
!!!!!
Success rate is 100 percent (5/5), round-trip min/avg/max = 12/29/56 ms
R3#ping ipv6 2001:da8:20e:2::1

Type escape sequence to abort.
Sending 5, 100-byte ICMP Echos to 2001:DA8:20E:2::1, timeout is 2 seconds:
!!!!!
Success rate is 100 percent (5/5), round-trip min/avg/max = 8/18/52 ms
```

可见，整个网络的连通性正常。

（2）查看 IPv6 的接口信息

```
R1#show ipv6 interface s1/0
Serial1/0 is up, line protocol is up
  IPv6 is enabled, link-local address is FE80::C800:12FF:FE88:0
                  //在 s1/0 上配了 IPv6 地址后，在 s1/0 上自动配置了本地链路地址：
                  //FE80::C800:12FF:FE88:0
  Global unicast address(es):
```

                2001:DA8:20E:5::1, subnet is 2001:DA8:20E:5::/64
                    //本接口的 IPv6 地址及子网
            Joined group address(es):   //加入的组播地址有（以下地址）：
                FF02::1   //表示本地链路上的所有节点和路由器
                FF02::2   //表示本地链路上的所有路由器
                FF02::9   //使用 FF02::9 地址作为目的更新地址
                FF02::1:FF00:1   //用于替换 ARP 机制的被请求节点的多播地址
                FF02::1:FF88:0   //与单播地址相关的被请求节点的多播地址
            MTU is 1500 bytes
            ICMP error messages limited to one every 100 milliseconds
            ICMP redirects are enabled
            ND DAD is enabled, number of DAD attempts: 1
            ND reachable time is 30000 milliseconds
            Hosts use stateless autoconfig for addresses.
    （3）查看路由表
        R1#show ipv6 route rip
        IPv6 Routing Table - 11 entries
        Codes: C - Connected, L - Local, S - Static, R - RIP, B - BGP
               U - Per-user Static route
               I1 - ISIS L1, I2 - ISIS L2, IA - ISIS interarea, IS - ISIS summary
               O - OSPF intra, OI - OSPF inter, OE1 - OSPF ext 1, OE2 - OSPF ext 2
               ON1 - OSPF NSSA ext 1, ON2 - OSPF NSSA ext 2
        R   2001:DA8:20E:3::/64 [120/3]
               via FE80::C800:16FF:FEC4:0, Serial1/0
        R   2001:DA8:20E:4::/64 [120/3]
               via FE80::C800:16FF:FEC4:0, Serial1/0
        R   2001:DA8:20E:6::/64 [120/2]
               via FE80::C800:16FF:FEC4:0, Serial1/0

可见，RIPng 的跳数与 RIP 相比，多了一跳，在 RIPng 中，默认情况下，进入路由表之前 RIPng 度量值加 1；RIPng 的下一跳不是邻居接口的 IP 地址，而是邻居路由器的链路本地地址，RIPng 使用链路本地地址作为更新消息的源地址，因此下一跳也使用邻居路由器的链路本地地址，此地址也可以通过"show ipv6 rip next-hops"来查看：

        R1#show ipv6 rip next-hops
        RIP process "AAA", Next Hops   //RIPng 的进程 AAA，下一跳地址为：
            FE80::C800:16FF:FEC4:0/Serial1/0 [4 paths]

下一跳地址是邻居路由器的链路本地地址，下面查看 R2 的 s1/0 接口信息，来证实"FE80::C800:16FF:FEC4:0"就是链路本地地址：

        R2#show ipv6 interface s1/0
        Serial1/0 is up, line protocol is up
            IPv6 is enabled, link-local address is FE80::C800:16FF:FEC4:0

可见，S1/0 的链路本地地址就是 FE80::C800:16FF:FEC4:0
            Global unicast address(es):
                2001:DA8:20E:5::2, subnet is 2001:DA8:20E:5::/64
            Joined group address(es):
                FF02::1

```
            FF02::2
            FF02::9
            FF02::1:FF00:2
            FF02::1:FFC4:0
    MTU is 1500 bytes
    ICMP error messages limited to one every 100 milliseconds
    ICMP redirects are enabled
    ND DAD is enabled, number of DAD attempts: 1
    ND reachable time is 30000 milliseconds
    Hosts use stateless autoconfig for addresses.
```

（4）查看 IPv6 RIPng 的数据库

```
R1#show ipv6 rip database
RIP process "AAA", local RIB
 2001:DA8:20E:3::/64, metric 3, installed
     Serial1/0/FE80::C800:16FF:FEC4:0, expires in 179 secs
 2001:DA8:20E:4::/64, metric 3, installed
     Serial1/0/FE80::C800:16FF:FEC4:0, expires in 179 secs
 2001:DA8:20E:5::/64, metric 2
     Serial1/0/FE80::C800:16FF:FEC4:0, expires in 179 secs
 2001:DA8:20E:6::/64, metric 2, installed
     Serial1/0/FE80::C800:16FF:FEC4:0, expires in 179 secs
```

这就是 R1 的 RIPng 数据库，其中 expires in 179 secs 表示还有 179s 路由条目过期。

（5）查看 RIPng 的相关信息

```
R1#show ipv6 RIP
RIP process "AAA", port 521, multicast-group FF02::9, pid 128
           //进程名为 AAA，UDP 端口号为 521，组播更新地址为 FF02::9，进程 id 为 128
     Administrative distance is 120. Maximum paths is 16
           //管理距离为 120，默认最大等价路径为 16 条
     Updates every 30 seconds, expire after 180
           //更新周期为 30s，过期周期为 180s
     Holddown lasts 0 seconds, garbage collect after 120
     Split horizon is on; poison reverse is off    //水平分割开启，毒化反转关闭
     Default routes are not generated
     Periodic updates 234, trigger updates 9
 Interfaces:
   Serial1/0
   Loopback0                  //在 S1/0 和 L0 上启用了 RIPng
 Redistribution:
   None                       //无重分布
```

实验 33　IPv6–over–IPv4 隧道

一、实验要求

- 掌握 IPv4 与 IPv6 共存方法；
- 掌握 IPv6-over-IPv4 隧道的配置方法。

二、实验说明

IPv6 的过渡策略主要包括三种：双栈（Dual Stacking）、隧道（Tunneling）和网络地址转换/协议转换技术（NAT-PT），其中双栈和隧道是使用最多的技术，能使用双栈的环境就可以使用隧道，下文中讲述了 IPv6-over-IPv4 隧道的配置方法。

IPv6-over-IPv4 隧道技术可以通过现有的运行 IPv4 协议的 Internet 骨干网络将局部的 IPv6 网络连接起来，采用隧道技术是 IPv4 向 IPv6 过渡初期最易于采用的技术。

三、实验拓扑

IPv6-over-IPv4 隧道配置使用的拓扑图如图 33-1 所示。

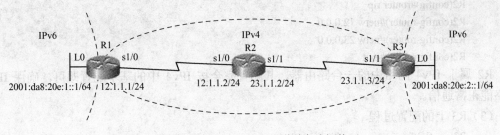

图 33-1　IPv6-over-IPv4 隧道实验拓扑

四、实验过程与实验总结

（1）R1 上的配置过程

```
R1(config)#ipv6 unicast-routing
R1(config)#int L0
R1(config-if)#ipv6 address 2001:da8:20e:1::1/64
                //配置 IPv6 地址，相当于这一边属于 IPv6 网络
R1(config-if)#int s1/0
```

```
R1(config-if)#ip add 12.1.1.1 255.255.255.0
                        //IPv6 网络中间穿过 IPv4 网络
R1(config-if)#no shut
R1(config-if)#exit
R1(config)#router rip
R1(config-router)#netw 12.0.0.0 //需要确保 IPv4 网络正常通信，否则隧道建立不起来
R1(config-router)#exit
R1(config)#int tunnel 0        //创建隧道端口
R1(config-if)#tunnel source s1/0 //指定隧道源端口为 s1/0，也可以是 s1/0 的 IP 地址
R1(config-if)#tunnel destination 23.1.1.3
                        //指定隧道的目标地址，在本例中，是 R3 的 s1/1 的地址，因为不
                        //是本地路由器，所以不能写成 s1/1，只能配置地址
R1(config-if)#tunnel mode ipv6ip //配置隧道模式为 IPv6-over-IPv4
R1(config-if)#ipv6 address 2001:da8:20e:10::1/64
                        //给隧道的一端配置一个 IPv6 地址
R1(config-if)#exit
R1(config)#ipv6 route ::/0 2001:da8:20e:10::2
                        //配置 IPv6 静态路由，这里指定的下一跳地址是隧道另一
                        //端的 IPv6 地址（也可以配置动态路由）
```

(2) R2 上的配置过程

```
R2(config)#int s1/0
R2(config-if)#ip add 12.1.1.2 255.255.255.0
R2(config-if)#no shut
R2(config-if)#int s1/1
R2(config-if)#ip address 23.1.1.1 255.255.255.0
R2(config-if)#no shut
R2(config-if)#exit
R2(config)#router rip
R2(config-router)#netw 12.0.0.0
R2(config-router)#netw 23.0.0.0
R2(config-router)#exit
```

R2 属于 IPv4 网络中的一台路由器，其配置完全按 IPv4 中的基本配置即可，确保 IPv4 网络能正常通信。

(3) R3 上的配置过程

```
R3(config)#ipv6 unicast-routing
R3(config)#int L0
R3(config-if)#ipv6 address 2001:da8:20e:2::1/64
R3(config-if)#int s1/1
R3(config-if)#ip add 23.1.1.3 255.255.255.0
R3(config-if)#no shut
R3(config-if)#exit
R3(config)#router rip
R3(config-router)#netw 23.0.0.0
R3(config-router)#exit
R3(config)#int tunnel 0
R3(config-if)#tunnel source s1/1
```

R3(config-if)#tunnel destination 12.1.1.1
R3(config-if)#tunnel mode ipv6ip
R3(config-if)#ipv6 address 2001:da8:20e:10::2/64
　　　　　　　　　　//给隧道的一端配置一个 IPv6 地址
R3(config-if)#exit
R3(config)#ipv6 route ::/0 2001:da8:20e:10::1
　　　　　　　　　　//配置 IPv6 静态路由，这里指定的下一跳地址是隧道另一
　　　　　　　　　　//端的 IPv6 地址（也可以配置动态路由）

配置完成后，在 R1 上以环回接口 L0 为原来 ping R3 上环回接口 L0 的 IPv6 地址：

R1#ping　ipv6　2001:da8:20e:2::1

Type escape sequence to abort.
Sending 5, 100-byte ICMP Echos to 2001:DA8:20E:2::1, timeout is 2 seconds:
!!!!!
Success rate is 100 percent (5/5), round-trip min/avg/max = 72/88/144ms
配置完成。

实验 34 基于子接口的帧中继配置

一、实验要求

- 掌握帧中继交换机的配置方法；
- 掌握点到点子接口帧中继的配置方法；
- 掌握多点子接口帧中继的配置方法。

二、实验说明

帧中继的子接口有两种类型：点到点（Point-to-Point）和多点子接口（Multipoint）。

点到点子接口是使用一个子接口来建立一条虚电路，该虚电路与远端路由器的一个子接口或物理接口相连，每个子接口都有单独的 DLCI（Data Link Connection Identifier，数据链路连接标识）号，并且各个子接口分属不同的网络。

多点子接口是使用一个子接口来建立多条虚电路，这些虚电路与远端路由器的多个子接口或物理接口相连，所有与该多点子接口相连的远端路由器子接口或物理接口属于同一个网络。

通常情况下，企业可使用多点子接口来连接远端的分支机构，而点对点子接口可在多路访问网络中模拟专线功能，但比专线费用更低，企业可用点对点子接口的帧中继建立与某些金融机构的连接，在这种方式的链路上不允许有第三方的接入点。

此实验在 DynamipsGUI 下完成，需要使用一台路由器模拟帧中继交换机。

三、实验拓扑

基于子接口的帧中继配置使用的拓扑如图 34-1 所示。

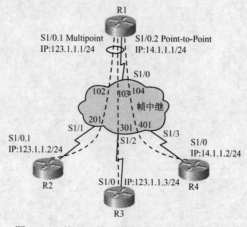

图 34-1 基于子接口的帧中继配置实验拓扑

四、实验过程

(1) 在帧中继交换机上的配置过程（由路由器模拟帧中继交换机）

  Router(config)#host FR
  FR(config)#frame-relay switching　　　　//在路由器上打开帧中继交换功能
  FR(config)#int s1/0
  FR(config-if)#encapsulation frame-relay //在 S1/0 端口中封装帧中继协议
  FR(config-if)#frame-relay intf-type dce　//将帧中继交换机上的端口设为 DCE 端
  FR(config-if)#frame-relay lmi-type ansi //指定 LMI 类型为 ANSI，如果不指定，默认是 Cisco
  FR(config-if)#clock rate 1200　　　　//配置时钟频率，由 DCE 端提供串行链路同步时钟
  FR(config-if)#frame-relay route 102 int s1/1 201
    //配置帧中继路由，告诉 R1 到达远端路由器 R2 的帧中继路径：帧中继交换机将从
    //S1/0 收到的 DLCI 号为 102 的数据帧向 S1/1 端口转发出去，并把 DLCI 号换成
    //201。在后面配置的 R1 上，有一条将 IP 地址为 192.168.0.2 与 DLCI 为 102 的静
    //态映射，在 R2 上，有一条将 192.168.0.1 与 DLCI 为 201 的静态映射，此处的指
    //令就是告诉从 R1 的 192.168.0.1 到 R2 的 192.168.0.2 该如何到达的路径
  FR(config-if)#frame-relay route 103 int s1/2 301
  FR(config-if)#frame-relay route 104 int s1/3 401
  FR(config-if)#no shut
  FR(config-if)#exit
  FR(config)#int s1/1　　　　　　　　//后面 s1/1、s1/2 和 s1/3 上的配置功能与 s1/0 类似
  FR(config-if)#encapsulation frame-relay
  FR(config-if)#frame-relay intf-type dce
  FR(config-if)#frame-relay lmi-type ansi
  FR(config-if)#clock rate 1200
  FR(config-if)#frame-relay route 201 interface s1/0 102
  FR(config-if)#no shut
  FR(config-if)#exit
  FR(config)#int s1/2
  FR(config-if)#encapsulation frame-relay
  FR(config-if)#frame-relay intf-type dce
  FR(config-if)#frame-relay lmi-type ansi
  FR(config-if)#clock rate 1200
  FR(config-if)#frame-relay route 301 int s1/0 103
  FR(config-if)#no shut
  FR(config-if)#int s1/3
  FR(config-if)#encapsulation frame-relay
  FR(config-if)#frame-relay intf-type dce
  FR(config-if)#frame-relay lmi-type ansi
  FR(config-if)#clock rate 1200
  FR(config-if)#frame-relay route 401 interface s1/0 104
  FR(config-if)#no shut

(2) 在路由器 R1 上的配置过程

  Router(config)#host R1
  R1(config)#int s1/0

```
R1(config-if)#encapsulation frame-relay    //在物理接口上封装帧中继，而非子接口
R1(config-if)#no shut                      //打开物理端口
R1(config-if)#exit
R1(config)#int s1/0.1 multipoint           //建立多点帧中继子接口
R1(config-subif)#ip address 123.1.1.1 255.255.255.0
                                           //为多点子接口配 IP 地址
R1(config-subif)#frame-relay map ip 123.1.1.2 102 broadcast
                                           //配置到达 R2 的帧中继映射，多点子接口的映射方
                                           //法和物理接口一样
R1(config-subif)#frame-relay map ip 123.1.1.3 103 broadcast
                                           //配置到达 R3 的帧中继映射
R1(config-subif)#exit
R1(config)#interface s1/0.2 point-to-point //建立点对点帧中继子接口
R1(config-subif)#ip address 14.1.1.1 255.255.255.0
R1(config-subif)#frame-relay interface-dlci 104
```

因为点对点帧中继链路上只有两个点，所以点对点子接口不需要映射 IP 地址到 DLCI，只需申明点对点子接口使用的 DLCI 号；而在物理接口帧中继和多点帧中继中，需要使用映射指令明确到哪个远程终端的连接。

（3）路由器 R2 上的配置过程
```
Router(config)#host R2
R2(config)#int s1/0
R2(config-if)#encapsulation frame-relay
R2(config-if)#no shut
R2(config-if)#exit
R2(config)#int s1/0.1 point-to-point
R2(config-subif)#ip add 123.1.1.2 255.255.255.0
R2(config-subif)#frame-relay interface-dlci 201
R2(config-fr-dlci)#exit
```

（4）路由器 R3 上的配置过程
```
Router(config)#host R3
R3(config)#int s1/0
R3(config-if)#encapsulation frame-relay
R3(config-if)#ip add 123.1.1.3 255.255.255.0
R3(config-if)#frame-relay map ip 123.1.1.1 301 broadcast
R3(config-if)#no shut
R3(config-if)#exit
```

（5）路由器 R4 上的配置过程
```
Router(config)#host R4
R4(config)#int s1/0
R4(config-if)#encapsulation frame-relay
R4(config-if)#ip add 14.1.1.2 255.255.255.0
R4(config-if)#frame-relay map ip 14.1.1.1 401 broadcast
R3(config-if)#no shut
R4(config-if)#exit
```

五、实验总结

配置完成后，查看到各远端路由器间虚电路的状态：

R1#show frame-relay pvc

......

DLCI = 102, DLCI USAGE = LOCAL, PVC STATUS = ACTIVE, INTERFACE = Serial1/0.1

......

DLCI = 103, DLCI USAGE = LOCAL, PVC STATUS = ACTIVE, INTERFACE = Serial1/0.1

......

DLCI = 104, DLCI USAGE = LOCAL, PVC STATUS = ACTIVE, INTERFACE = Serial1/0.2

......

可见，DLCI 号为 102、103、104 的三条 PVC 均是 ACTIVE 状态，是正常状态。再查看路由器 R1 通过虚电路到达远端路由器的 DLCI 与 IP 地址间的映射情况：

R1#show frame-relay map

Serial1/0.1 (up): ip 123.1.1.3 dlci 103(0x67,0x1870), static,
        broadcast,
        CISCO, status defined, active

Serial1/0.1 (up): ip 123.1.1.2 dlci 102(0x66,0x1860), static,
        broadcast,
        CISCO, status defined, active

Serial1/0.2 (up): point-to-point dlci, dlci 104(0x68,0x1880), broadcast
        status defined, active

可见，前两条映射是使用的帧中继多点子接口，后面一条是使用的点到点映射状态。最后来验证路由器 R1 与远端路由器 R2、R3、R4 的连通性，均可正常通信。

实验 35 帧中继的逆向 ARP

一、实验要求

- 了解逆向 ARP 的作用；
- 掌握逆向 ARP 的配置方法。

二、实验说明

在帧中继中配置客户端路由器时，使用了"frame-relay map ip"指令，将本地的 DLCI 号与远端的 IP 地址建立起静态映射，这有点类似于静态路由，两者都采用手工配置，如果在复杂的帧中继环境中，会增大管理复杂程度，而且不适应帧中继拓扑的变化。而逆向 ARP 是用户端路由器将本地的 DLCI 号去映射远端设备的三层 IP 地址，从而建立起帧中继映射。

三、实验拓扑

本实验的拓扑图如图 35-1 所示。

图 35-1 逆向 ARP 实验拓扑

四、配置过程

（1）帧中继交换机的配置过程

```
Router(config)#host FR
FR(config)#frame-relay switching
FR(config)#int s1/0
FR(config-if)#encapsulation frame-relay
FR(config-if)#frame-relay intf-type dce
FR(config-if)#clock rate 1200
FR(config-if)#frame-relay route 102 int s1/1 201
FR(config-if)#no shut
FR(config-if)#int s1/1
FR(config-if)#encapsulation frame-relay
```

```
FR(config-if)#frame-relay intf-type dce
FR(config-if)#clock rate 1200
FR(config-if)#frame-relay route 201 int s1/0 102
FR(config-if)#no shut
FR(config-if)#exit
```
（2）路由器 R1 的配置过程
```
Router(config)#host R1
R1(config)#int s1/0
R1(config-if)#encapsulation frame-relay
R1(config-if)#ip add 192.168.1.1 255.255.255.0
R1(config-if)#no shut
```
（3）路由器 R2 的配置过程
```
Router(config)#host R2
R2(config)#int s1/0
R2(config-if)#encapsulation frame-relay
R2(config-if)#ip add 192.168.1.2 255.255.255.0
R2(config-if)#no shut
```

五、实验总结

在配置路由器 R1、R2 时，不再需要使用 frame-relay map ip 命令来配置静态映射了，只需要配置帧中继封装和 IP 地址即可。配置完成后，查看帧中继的映射情况：

```
R1#show frame-relay map
Serial1/0 (up): ip 192.168.1.2 dlci 102(0x66,0x1860), dynamic,
                broadcast,, status defined, active
```

可见，在路由器 R1 上建立了一条从本地 DLCI 为 102 到远端 IP 192.168.1.2 的 dynamic（动态）映射。这与前面的静态映射是不一样的。

再进行连通性测试：

```
R1#ping 192.168.1.2

Type escape sequence to abort.
Sending 5, 100-byte ICMP Echos to 192.168.1.2, timeout is 2 seconds:
!!!!!
Success rate is 100 percent (5/5), round-trip min/avg/max = 52/128/272 ms
```

可见，路由器 R1 与 R2 通过动态映射的帧中继能正常通信。

帧中继的逆向 ARP 功能是 Cisco 设备默认启用的，但如果在路由器 R1、R2 上配置了静态映射，则逆向 ARP 解析将自动关闭。另外，也可用手动方式关闭逆向 ARP 解析。如果手动关闭后，则必须配置静态映射：

```
R1(config)#int s1/0
R1(config-if)#no frame-relay inverse-arp
R1(config-if)#frame-relay map ip 192.168.1.2 102 broadcast
```

实验 36　RIP Over 帧中继

一、实验要求

- 掌握在帧中继网络上配置动态路由协议 RIP 的方法；
- 理解在帧中继网络中多点子接口和点对点子接口上水平分割的区别；
- 掌握在帧中继中用户端路由器 ping 本端不通的解决方法。

二、实验说明

由于 RIP 存在水平分割，会导致半网状帧中继的 RIP 路由学习不正常，在多点子接口和物理接口上，都存在水平分割的问题。如果在全网状帧中继中，由于路由器 R2 与 R3 间的路由更新不需要经过 R1 的 S1/0 接口转发，所以不会存在水平分割的问题，本实验讲的是在半网状帧中继上的 RIP 配置。

三、实验拓扑

本实验的拓扑图如图 36-1 所示。

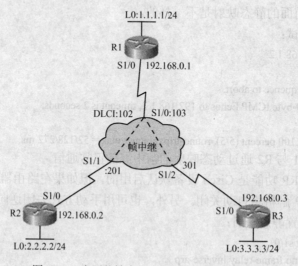

图 36-1　半网状帧中继上的 RIP 配置实验拓扑

四、实验过程

帧中继交换机的配置与实验 34 中的帧中继交换机的配置一样，在此略。

（1）路由器 R1 的配置过程

  Router(config)#host R1
  R1(config)#int s1/0
  R1(config-if)#encapsulation frame-relay
  R1(config-if)#ip add 192.168.0.1 255.255.255.0
  R1(config-if)#no shut
  R1(config-if)#frame-relay map ip 192.168.0.2 102 broadcast
  R1(config-if)#frame-relay map ip 192.168.0.3 103 broadcast
  R1(config-if)#int l0
  R1(config-if)#ip add 1.1.1.1 255.255.255.0
  R1(config-if)#router rip   //配置 RIP 协议
  R1(config-router)#no auto-summary
  R1(config-router)#version 2
  R1(config-router)#netw 192.168.0.0   //公告网络
  R1(config-router)#netw 1.1.1.0
  R1(config-router)#exit

（2）路由器 R2 的配置过程

  Router(config)#host R2
  R2(config)#int s1/0
  R2(config-if)#encapsulation frame-relay
  R2(config-if)#ip add 192.168.0.2 255.255.255.0
  R2(config-if)#no shut
  R2(config-if)#frame-relay map ip 192.168.0.1 201 broadcast
  R2(config-if)#frame-relay map ip 192.168.0.3 201 broadcast
  R2(config-if)#int l0
  R2(config-if)#ip add 2.2.2.2 255.255.255.0
  R2(config-if)#router rip
  R2(config-router)#ver 2
  R2(config-router)#no auto-summary
  R2(config-router)#netw 192.168.0.0
  R2(config-router)#net 2.2.2.0

（3）路由器 R3 的配置过程

  Router(config)#host R3
  R3(config)#int s1/0
  R3(config-if)#ip add 192.168.0.3 255.255.255.0
  R3(config-if)#no shut
  R3(config-if)#encapsulation frame-relay
  R3(config-if)#frame-relay map ip 192.168.0.1 301 broadcast
  R3(config-if)#frame-relay map ip 192.168.0.2 301 broadcast
  R3(config-if)#int l0
  R3(config-if)#ip add 3.3.3.3 255.255.255.0
  R3(config-if)#router rip

```
R3(config-router)#ver 2
R3(config-router)#no auto-summary
R3(config-router)#netw 192.168.0.0
R3(config-router)#netw 3.3.3.0
```

五、实验总结

配置完成后，查看路由器的路由表：

```
R1#show ip route
……
     1.0.0.0/24 is subnetted, 1 subnets
C       1.1.1.0 is directly connected, Loopback0
     2.0.0.0/24 is subnetted, 1 subnets
R       2.2.2.0 [120/1] via 192.168.0.2, 00:00:15, Serial1/0
     3.0.0.0/24 is subnetted, 1 subnets
R       3.3.3.0 [120/1] via 192.168.0.3, 00:00:24, Serial1/0
C    192.168.0.0/24 is directly connected, Serial1/0
R2#show ip route
……
     1.0.0.0/24 is subnetted, 1 subnets
R       1.1.1.0 [120/1] via 192.168.0.1, 00:00:22, Serial1/0
     2.0.0.0/24 is subnetted, 1 subnets
C       2.2.2.0 is directly connected, Loopback0
     3.0.0.0/24 is subnetted, 1 subnets
R       3.3.3.0 [120/2] via 192.168.0.3, 00:00:22, Serial1/0
C    192.168.0.0/24 is directly connected, Serial1/0
R3#show ip route
……
     1.0.0.0/24 is subnetted, 1 subnets
R       1.1.1.0 [120/1] via 192.168.0.1, 00:00:23, Serial1/0
     2.0.0.0/24 is subnetted, 1 subnets
R       2.2.2.0 [120/2] via 192.168.0.2, 00:00:23, Serial1/0
     3.0.0.0/24 is subnetted, 1 subnets
C       3.3.3.0 is directly connected, Loopback0
C    192.168.0.0/24 is directly connected, Serial1/0
```

从路由器 R1、R2、R3 的路由表输出可见，各路由器的路由学习都是正常的，并没有出现前面提到的由于水平分割而导致的半网状帧中继的 RIP 路由学习不正常的情况，这是因为帧中继物理接口默认关闭了水平分割。可以通过查看 R1 的 S1/0 端口来证实：

```
R1#show ip int s1/0
Serial1/0 is up, line protocol is up
  Internet address is 192.168.0.1/24
  Broadcast address is 255.255.255.255
  Address determined by non-volatile memory
  MTU is 1500 bytes
  Helper address is not set
  Directed broadcast forwarding is disabled
  Multicast reserved groups joined: 224.0.0.9
```

```
        Outgoing access list is not set
        Inbound   access list is not set
        Proxy ARP is enabled
        Local Proxy ARP is disabled
        Security level is default
        Split horizon is disabled              //可见，默认情况下水平分割是关闭的
```
如果在 R1 的 S1/0 上开启水平分割：
```
    R1(config)#int s1/0
    R1(config-if)#ip split-horizon
```
然后在 R2、R3 上查看路由表（由于 RIP 收敛较慢，要么过一会再查看，要么运行"R2#clear ip route *"命令后查看）：
```
    R2#show ip route
    ……
         1.0.0.0/24 is subnetted, 1 subnets
    R       1.1.1.0 [120/1] via 192.168.0.1, 00:00:14, Serial1/0
         2.0.0.0/24 is subnetted, 1 subnets
    C       2.2.2.0 is directly connected, Loopback0
    C    192.168.0.0/24 is directly connected, Serial1/0
    R3#show ip route
    ……
         1.0.0.0/24 is subnetted, 1 subnets
    R       1.1.1.0 [120/1] via 192.168.0.1, 00:00:01, Serial1/0
         3.0.0.0/24 is subnetted, 1 subnets
    C       3.3.3.0 is directly connected, Loopback0
    C    192.168.0.0/24 is directly connected, Serial1/0
```
可见，水平分割已产生作用，R1 不再把从 S1/0 端口学到的 R2 上环回接口的路由再发给 R3，也不再把学到的 R3 上环回接口的路由发给 R2。解决帧中继的水平分割问题，一是可以关闭水平分割，二是使用点到点子接口的帧中继。

与物理接口的帧中继默认水平分割是关闭的不同，在多点子接口帧中继上，水平分割默认是开启的。下面以图 36-2 为拓扑来证实多点子接口帧中继的水平分割。

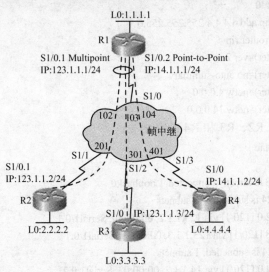

图 36-2 多点子接口帧中继的水平分割拓扑

在路由器 R1、R2、R3、R4 上，均配置环回接口，并公告 RIP 协议（这里仅配置相对于实验 34 的新增配置）：

- 在 R1 上新增的配置：

    R1(config)#int l0
    R1(config-if)#ip add 1.1.1.1 255.0.0.0
    R1(config-if)#router rip
    R1(config-router)#version 2
    R1(config-router)#no auto-summary
    R1(config-router)#netw 123.1.1.0
    R1(config-router)#net 14.0.0.0
    R1(config-router)#netw 1.0.0.0

- 在 R2 上新增的配置：

    R2(config)#int l0
    R2(config-if)#ip add 2.2.2.2 255.255.255.0
    R2(config-if)#router rip
    R2(config-router)#ver 2
    R2(config-router)#no auto-summary
    R2(config-router)#netw 123.0.0.0
    R2(config-router)#netw 2.0.0.0

- 在 R3 上新增的配置：

    R3(config)#int l0
    R3(config-if)#ip add 3.3.3.3 255.0.0.0
    R3(config-if)#router rip
    R3(config-router)#ver 2
    R3(config-router)#no auto-summary
    R3(config-router)#net 3.0.0.0
    R3(config-router)#netw 123.0.0.0

- 在 R4 上新增的配置：

    R4(config)#int l0
    R4(config-if)#ip add 4.4.4.4 255.255.255.0
    R4(config-if)#router rip
    R4(config-router)#ver 2
    R4(config-router)#no auto-summary
    R4(config-router)#netw 4.0.0.0
    R4(config-router)#netw 14.0.0.0

然后，查看 R1、R2、R3 和 R4 的路由表：

    R1#show ip route
    ......
    C   1.0.0.0/8 is directly connected, Loopback0
            2.0.0.0/24 is subnetted, 1 subnets
    R   2.2.2.0 [120/1] via 123.1.1.2, 00:00:10, Serial1/0.1
    R   3.0.0.0/8 [120/1] via 123.1.1.3, 00:00:19, Serial1/0.1
            4.0.0.0/24 is subnetted, 1 subnets
    R   4.4.4.0 [120/1] via 14.1.1.2, 00:00:03, Serial1/0.2
            123.0.0.0/24 is subnetted, 1 subnets

```
        C        123.1.1.0 is directly connected, Serial1/0.1
                 14.0.0.0/24 is subnetted, 1 subnets
        C        14.1.1.0 is directly connected, Serial1/0.2
```

可见，R1 学到了所有网络的路由条目，并且 ping R2、R3、R4 上的所有 IP 地址，均可 ping 通。

```
R4#show ip route
……
        R        1.0.0.0/8 [120/1] via 14.1.1.1, 00:00:22, Serial1/0
                 2.0.0.0/24 is subnetted, 1 subnets
        R        2.2.2.0 [120/2] via 14.1.1.1, 00:00:22, Serial1/0
        R        3.0.0.0/8 [120/2] via 14.1.1.1, 00:00:22, Serial1/0
                 4.0.0.0/24 is subnetted, 1 subnets
        C        4.4.4.0 is directly connected, Loopback0
                 123.0.0.0/24 is subnetted, 1 subnets
        R        123.1.1.0 [120/1] via 14.1.1.1, 00:00:22, Serial1/0
                 14.0.0.0/24 is subnetted, 1 subnets
        C        14.1.1.0 is directly connected, Serial1/0
```

可见，由于点对点子接口帧中继不存在水平分割问题，所以 R4 也学到了所有网络的路由条目，并且 ping R2、R3、R4 上的所有 IP 地址，也均可 ping 通。

```
R2#show ip route
……
        R        1.0.0.0/8 [120/1] via 123.1.1.1, 00:00:25, Serial1/0.1
                 2.0.0.0/24 is subnetted, 1 subnets
        C        2.2.2.0 is directly connected, Loopback0
                 4.0.0.0/24 is subnetted, 1 subnets
        R        4.4.4.0 [120/2] via 123.1.1.1, 00:00:25, Serial1/0.1
                 123.0.0.0/24 is subnetted, 1 subnets
        C        123.1.1.0 is directly connected, Serial1/0.1
                 14.0.0.0/24 is subnetted, 1 subnets
        R        14.1.1.0 [120/1] via 123.1.1.1, 00:00:25, Serial1/0.1

R3#show ip route
……
        R        1.0.0.0/8 [120/1] via 123.1.1.1, 00:00:09, Serial1/0
        C        3.0.0.0/8 is directly connected, Loopback0
                 4.0.0.0/24 is subnetted, 1 subnets
        R        4.4.4.0 [120/2] via 123.1.1.1, 00:00:09, Serial1/0
                 123.0.0.0/24 is subnetted, 1 subnets
        C        123.1.1.0 is directly connected, Serial1/0
                 14.0.0.0/24 is subnetted, 1 subnets
        R        14.1.1.0 [120/1] via 123.1.1.1, 00:00:09, Serial1/0
```

可见，R2 没有学到 R3 上 L0 接口的路由，同时，R3 也没学到 R2 上 L0 接口的路由。

其原因是在多点子接口帧中继上，水平分割默认是开启的。要使 RIP 运行在多点子接口上不受影响，需要在子接口模式下，关闭水平分割：

```
R1(config-if)#int s1/0.1
R1(config-subif)#no ip split-horizon
```

然后再去 R2（或 R3）上查看路由表：

R2#show ip route

……

R 1.0.0.0/8 [120/1] via 123.1.1.1, 00:00:19, Serial1/0.1
   2.0.0.0/24 is subnetted, 1 subnets
C 2.2.2.0 is directly connected, Loopback0
R 3.0.0.0/8 [120/2] via 123.1.1.3, 00:00:19, Serial1/0.1
   4.0.0.0/24 is subnetted, 1 subnets
R 4.4.4.0 [120/2] via 123.1.1.1, 00:00:19, Serial1/0.1
   123.0.0.0/24 is subnetted, 1 subnets
C 123.1.1.0 is directly connected, Serial1/0.1
   14.0.0.0/24 is subnetted, 1 subnets
R 14.1.1.0 [120/1] via 123.1.1.1, 00:00:19, Serial1/0.1

可见，R2 已经学习到来自 R3 的环回接口的路由条目。

最后，再讲一个在帧中继中用户端路由器 ping 本端不通的问题，在前面讲的所有帧中继配置中，在测试连通性时，都是测试的两用户端路由器间的连通性，如果是路由器 ping 本端呢，结果会怎么样？以图 36-1 中的 R1 为例来测试：

R1#ping 192.168.0.1

Type escape sequence to abort.
Sending 5, 100-byte ICMP Echos to 192.168.0.1, timeout is 2 seconds:
.....
Success rate is 0 percent (0/5)

发现 ping 本端不通，其原因在本端的 DLCI 号与本端的 IP 地址间没有建立映射关系，所以需要增加一条配置：

R1(config)#int s1/0
R1(config-if)#frame-relay map ip 192.168.0.1 102 broadcast
     //将本端的 DLCI 号与本端的 IP 地址建立静态映射
R1(config-if)#end

然后再 ping 本端：

R1#ping 192.168.0.1

Type escape sequence to abort.
Sending 5, 100-byte ICMP Echos to 192.168.0.1, timeout is 2 seconds:
!!!!!
Success rate is 100 percent (5/5), round-trip min/avg/max = 36/66/156 ms

可见，此时可 ping 通本端了。

实验 37　在多点子接口帧中继下运行 OSPF

一、实验要求

- 掌握多点子接口帧中继的配置方法；
- 掌握 NBMA 模式下 OSPF 的配置方法；
- 掌握手工指定 OSPF 邻居。

二、实验说明

在本实验中，虽然讲的是多点子接口帧中继配置 OSPF 的过程，但对物理接口来说，配置方法和多点子接口帧中继是一样的，而基于点对点子接口的帧中继配置方法与此不同，在实验 34 中已讲述了基于点对点子接口帧中继的配置方法。

三、实验拓扑

本实验的拓扑图如图 37-1 所示。

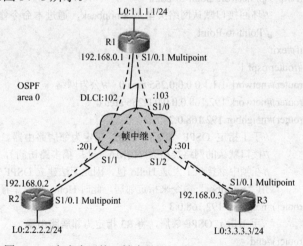

图 37-1　在多点子接口帧中继下运行 OSPF 实验拓扑

四、实验过程

帧中继交换机的配置与实验 34 中的帧中继交换机的配置一样，在此略。

(1) 路由器 R1 的配置过程

```
Router>en
Router#conf t
Router(config)#host R1
R1(config)#line console 0
R1(config-line)#exec-timeout 0 0          //配置 console 线路永不超时
R1(config-line)#end
R1#conf t
R1(config)#int s1/0
R1(config-if)#encapsulation frame-relay   //在物理接口上封装帧中继，而非子接
                                            口上
R1(config-if)#no frame-relay inverse-arp  //关闭帧中继逆向 arp 解析
R1(config-if)#no shut                     //打开物理端口，在子接口下不需再打开
R1(config)#interface serial 1/0.1 multipoint //建立多点帧中继子接口
R1(config-subif)#ip add 192.168.0.1 255.255.255.0
                          //为多点子接口配置 IP 地址
R1(config-subif)#frame-relay map ip 192.168.0.2 102 broadcast
                //配置到达 R2 的帧中继映射，多点子接口的映射方法和物理接口一样
R1(config-subif)#frame-relay map ip 192.168.0.3 103 broadcast
                //配置到达 R3 的帧中继映射
R1(config-subif)#frame-relay map ip 192.168.0.1 102 broadcast
                //将本端的 DLCI 号与本端的 IP 地址建立静态映射，使之可 ping 通本端
R1(config-subif)#no frame-relay inverse-arp
                //关闭帧中继逆向 arp 解析
R1(config)#int L0
R1(config-if)#ip add 1.1.1.1 255.255.255.0     //配置环回接口 IP，用于测试 OSPF 配置
R1(config-if)#ip ospf network point-to-point
                //环回接口默认网络类型是 Loopback，通过本命令修改普通接口的网络类型为
                //Point-to-Point
R1(config-if)#exit
R1(config)#router ospf 1
R1(config-router)#network 1.1.1.0 0.0.0.255 area 0   //公告网络
R1(config-router)#network 192.168.0.0 0.0.0.255 area 0
R1(config-router)#neighbor 192.168.0.2
                //手工指定 OSPF 邻居，将 R2 指定为邻居路由器。在帧中继网络上，OSPF、
                //接口默认的网络类型为 NBMA（非广播多路访问），在这种模式下 OSPF 不会
                //在帧中继接口上发送 Hello 包，因此无法建立 OSPF 的邻接关系。可以手动使//
                //用"neighbor"命令来指定邻居，此时 Hello 包以单播形式传送
R1(config-router)#neighbor 192.168.0.3
                //手工指定 OSPF 邻居，将 R3 指定为邻居路由器
R1(config-router)#end
R1#wr
```

(2) 路由器 R2 的配置过程

```
Router>en
Router#conf t
Router(config)#host R2
R2(config)#line console 0
```

```
R2(config-line)#exec-timeout 0 0
R2(config-line)#exit
R2(config)#int s1/0
R2(config-if)#encapsulation frame-relay
R2(config-if)#no frame-relay inverse-arp
R2(config-if)#no shut
R2(config)#int L0
R2(config-if)#ip add 2.2.2.2 255.255.255.0
R2(config-if)#ip ospf network point-to-point
R2(config-if)#exit
R2(config)#interface serial 1/0.1 multipoint
R2(config-subif)#ip add 192.168.0.2 255.255.255.0
R2(config-subif)#ip ospf priority 0
            //配置 OSPF 优先级为 0，使本路由器 R2 不参与 DR 的选举，从而使居于中心位置
            //的路由器 R1 为 DR，在 R3 上也是同样配置
R2(config-subif)#frame-relay map ip 192.168.0.1 201 broadcast
R2(config-subif)#frame-relay map ip 192.168.0.3 201 broadcast
R2(config-subif)#frame-relay map ip 192.168.0.2 201 broadcast
R2(config-subif)#no frame-relay inverse-arp
R2(config-subif)#router ospf 1
R2(config-router)#netw 192.168.0.0 0.0.0.255 area 0
R2(config-router)#netw 2.2.2.0 0.0.0.255 area 0
R2(config-router)#end
R2#wr
```

（3）路由器 R3 的配置过程

```
Router>en
Router#conf t
Router(config)#host R3
R3(config)#line console 0
R3(config-line)#exec-timeout 0 0
R3(config-line)#end
R3#conf t
R3(config)#int s1/0
R3(config-if)#encapsulation frame-relay
R3(config-if)#no frame-relay inverse-arp
R3(config-if)#no shut
R3(config)#interface serial 1/0.1 multipoint
R3(config-subif)#ip add 192.168.0.3 255.255.255.0
R3(config-subif)#ip ospf priority 0
R3(config-subif)#frame-relay map ip 192.168.0.1 301 broadcast
R3(config-subif)#frame-relay map ip 192.168.0.2 301 broadcast
R3(config-subif)#frame-relay map ip 192.168.0.3 301 broadcast
R3(config-subif)#no frame-relay inverse-arp
R3(config-subif)#exit
R3(config)#int L0
R3(config-if)#ip add 3.3.3.3 255.255.255.0
```

```
R3(config-if)#ip ospf network point-to-point
R3(config-if)#exit
R3(config)#router ospf 1
R3(config-router)#netw 192.168.0.0 0.0.0.255 area 0
R3(config-router)#netw 3.3.3.0 0.0.0.255 area 0
R3(config-router)#end
R3#wr
```

五、实验总结

（1）查看帧中继映射

```
R1#show frame-relay map
Serial1/0.1 (up): ip 192.168.0.1 dlci 102(0x66,0x1860), static,
          broadcast,
          CISCO, status defined, active          //指向自己的静态映射
Serial1/0.1 (up): ip 192.168.0.2 dlci 102(0x66,0x1860), static,
          broadcast,
          CISCO, status defined, active          //指向R2的静态映射
Serial1/0.1 (up): ip 192.168.0.3 dlci 103(0x67,0x1870), static,
          broadcast,
          CISCO, status defined, active          //指定R3的静态映射
```

可见，路由器 R1 的多点子接口上，建立有三条帧中继的静态映射，并可查看到达对端的 IP 地址，而在点到点帧中继则不能看到对端 IP 地址。

（2）查看虚电路状态

```
R1#show frame-relay pvc

PVC Statistics for interface Serial1/0 (Frame Relay DTE)

             Active      Inactive    Deleted     Static
Local        2           0           0           0
Switched     0           0           0           0
Unused       0           0           0           0

DLCI = 102, DLCI USAGE = LOCAL, PVC STATUS = ACTIVE, INTERFACE = Serial1/0.1
          //这条 PVC 的 DLCI 为 102，状态是 ACTIVE（活跃），所连接口是 Serial1/0.1
  input pkts 0              output pkts 0             in bytes 0
  out bytes 0               dropped pkts 0            in pkts dropped 0
  out pkts dropped 0                  out bytes dropped 0
  in FECN pkts 0            in BECN pkts 0            out FECN pkts 0
  out BECN pkts 0           in DE pkts 0              out DE pkts 0
  out bcast pkts 0          out bcast bytes 0
  5 minute input rate 0 bits/sec, 0 packets/sec
  5 minute output rate 0 bits/sec, 0 packets/sec
  pvc create time 00:02:16, last time pvc status changed 00:01:06
```

　　　　　　DLCI = 103, DLCI USAGE = LOCAL, PVC STATUS = ACTIVE, INTERFACE = Serial1/0.1

```
        input pkts 0           output pkts 0           in bytes 0
        out bytes 0            dropped pkts 0          in pkts dropped 0
         out pkts dropped 0              out bytes dropped 0
        in FECN pkts 0         in BECN pkts 0          out FECN pkts 0
        out BECN pkts 0        in DE pkts 0            out DE pkts 0
        out bcast pkts 0       out bcast bytes 0
        5 minute input rate 0 bits/sec, 0 packets/sec
        5 minute output rate 0 bits/sec, 0 packets/sec
        pvc create time 00:02:19, last time pvc status changed 00:01:08
```

可见，路由器 R1 上有两条虚电路，DLCI 号为 102 和 103，其状态（PVC STATUS）均为 ACTIVE，说明这两条虚电路为活跃状态，是正常状态。PVC 有三种状态：

- ACTIVE：活跃状态，帧中继链路工作正常。
- INACTIVE：非活跃状态，表示帧中继交换机本地连接正常，但远端路由器到帧中继交换机的链路有故障，如端口配置错误、未开启等。
- DELETED：表示本路由器没有从帧中继交换机收到关于帧中继的 LMI 信令，一般是本路由器配置有问题，如 DLCI 号配置错误。

（3）查看接口上的 OSPF 信息

```
    R1#show ip ospf interface s1/0.1
            //这里如果没有加 s1/0.1，则还会显示本路由器的 L0 接口上的 OSPF 信息
    Serial1/0.1 is up, line protocol is up      //物理层和线路协议均 up，表示工作正常
      Internet Address 192.168.0.1/24, Area 0   //接口 IP 和所在区域
      Process ID 1, Router ID 1.1.1.1, Network Type NON_BROADCAST, Cost: 64
                                //RID 是 L0 的 IP，接口的网络类型为 NBMA
      Transmit Delay is 1 sec, State DR, Priority 1
                //接口为 DR，OSPF 优先级为 1（前面在配置 R2 和 R3 的 OSPF 端口时，将优先
                //级配为 0，确保 R1 为 DR）
      Designated Router (ID) 1.1.1.1, Interface address 192.168.0.1
                //DR 的 ID 为 1.1.1.1，本接口 ID 为 192.168.0.1
      No backup designated router on this network       //无 BDR
      Timer intervals configured, Hello 30, Dead 120, Wait 120, Retransmit 5
        oob-resync timeout 120
        Hello due in 00:00:14
      Index 2/2, flood queue length 0
      Next 0x0(0)/0x0(0)
      Last flood scan length is 1, maximum is 1
      Last flood scan time is 0 msec, maximum is 4 msec
      Neighbor Count is 2, Adjacent neighbor count is 2
                //有两个 OSPF 邻居，形成邻接关系的数量有两个
        Adjacent with neighbor 2.2.2.2
        Adjacent with neighbor 3.3.3.3
                    //R1 与 R2、R3 均形成了邻接关系
      Suppress hello for 0 neighbor(s)
```

（4）查看路由表

```
R1#show ip route ospf
     2.0.0.0/24 is subnetted, 1 subnets
O       2.2.2.0 [110/65] via 192.168.0.2, 00:14:14, Serial1/0.1
     3.0.0.0/24 is subnetted, 1 subnets
O       3.3.3.0 [110/65] via 192.168.0.3, 00:14:14, Serial1/0.1
R2#show ip route ospf
     1.0.0.0/24 is subnetted, 1 subnets
O       1.1.1.0 [110/65] via 192.168.0.1, 00:17:12, Serial1/0.1
     3.0.0.0/24 is subnetted, 1 subnets
O       3.3.3.0 [110/65] via 192.168.0.3, 00:17:12, Serial1/0.1
     //到达 3.3.3.0/24 网络的下一跳地址是 192.168.0.3，是因为在 R2 上配置了一条到达
     //192.168.0.3 的映射：frame-relay map ip 192.168.0.3 201 broadcast，使得网络 3.3.3.0/24
     //就是通过 192.168.0.3 学到的
R3#show ip route ospf
     1.0.0.0/24 is subnetted, 1 subnets
O       1.1.1.0 [110/65] via 192.168.0.1, 00:17:36, Serial1/0.1
     2.0.0.0/24 is subnetted, 1 subnets
O       2.2.2.0 [110/65] via 192.168.0.2, 00:17:36, Serial1/0.1
```
路由器 R1、R2、R3 的 OSPF 路由表的输出，表明整个网络是互通的。

参 考 文 献

[1] Jeff Doyle. TCP/IP 路由技术（第一卷） [M]. 2 版. 葛建立，吴剑章，夏俊杰，译.北京：人民邮电出版社，2013.

[2] Jeff Doyle. TCP/IP 路由技术（第二卷）（全新翻译版） [M]. 葛建立，吴剑章，夏俊杰 ，译. 北京：人民邮电出版社，2013.

[3] Richard Deal. CCNA 学习指南[M]. 邢京武，何涛，译. 北京：人民邮电出版社，2006

[4] 崔北亮. CCNA 认证指南[M]. 北京：电子工业出版社，2009.

[5] 梁广民，王隆杰. 思科网络实验室 CCNA 实验指南[M]. 北京：电子工业出版社，2009.

[6] 杨靖，刘亮. 实用网络技术配置指南[M]. 北京：北京希望电子出版社，2006.

[7] 冯昊，等. 交换机/路由器的配置与管理[M]. 北京：清华大学出版社，2005

[8] 鲁士文. 计算机网络习题与解析[M]. 第 2 版. 北京：清华大学出版社，2005.

[9] 刘晓辉. 网络设备[M]. 北京：机械工业出版社，2007.

[10] 颜凯，杨宁，李育强，高翔. CCNP2 远程接入[M]. 北京：人民邮电出版社，2007.